Michael Aufhauser

Meine schönsten Hundegeschichten

A Happy Ending for Rescued Dogs

Michael Aufhauser

Meine schönsten Hundegeschichten

A Happy Ending for Rescued Dogs

teNeues

Inhalt | Content

GUT AIDERBICHL

Vorwort

Viele Hunde haben mich in den letzten Jahrzehnten durch mein Leben begleitet. Ein Mensch-Hund-Leben, das schöner nicht sein könnte. Trotzdem würde ich mich nicht unbedingt als Hundekenner bezeichnen. Denn jeder meiner Hunde ist anders, das war schon immer so. Vom vermeintlich aggressiven Schäferhund bis hin zum ängstlichen kleinen Mischling aus Spanien: keiner wie der andere.

Sie kamen alle aus demselben Grund zu mir: jeder dieser Hunde musste aus großer Not gerettet werden. Außerdem waren sie schwer oder gar nicht an andere gute Plätze zu vermitteln. Nur mein erster, ein ungarischer Vorstehhund, der Pirosch hieß, war nicht arm, aber für mich vielleicht der Wichtigste, weil mit ihm damals mein Umdenken begann.

Mit den schönsten Geschichten, die ich mit meinen Hunden erlebte, gebe ich auch einen Teil meines persönlichen Lebens und meiner Entwicklung preis. Spätestens als ich in Málaga zusehen musste, wie Hunde vergast wurden, war mir klar, dass sich alles ändern muss. Und ich mich selber auch.

Auf den 11 Höfen von Gut Aiderbichl in Deutschland und Österreich leben über 1000 gerettete Tiere. Darunter auch eine Vielzahl von Hunden, die bei uns ein neues und sicheres Zuhause gefunden haben.

Ich bitte um Nachsicht, dass sich in diesem Buch manchmal die Grenzen zwischen Herr und Hund verwischen. Schließlich berichte ich von Beziehungskisten der ganz besonderen Art. Jede einzelne Episode ist ein Beispiel dafür, wie ähnlich uns Tiere sind. Besonders in Sachen Liebe.

Preface

I've had a lot of dogs accompany me through life over the last decades. I'm talking about a man/dog existence that leaves nothing to be desired. Yet, I hesitate to think of myself as an expert on dogs. You see, none of my dogs are like the other, they never have been. Be it that purportedly aggressive German Shepherd or that timid little mongrel from Spain—each one is unique.

What they do have in common is the reason that brought them to me: All my dogs had to be rescued from some dire situation. Also, finding other good places for them proved either very hard or downright impossible. My first dog ever, a Hungarian pointer named Pirosch, had never been unfortunate at all. To me, however, he may be the most significant of them all, because it was this dog back in the day that began to change my perspective.

By sharing the most precious stories I've experienced with my dogs, I also reveal part of my personal life and development. To me, having to witness dogs being gassed to death in Málaga, Spain, was the final wake-up call that things had to change in a profound way. And so did I.

The 11 sanctuaries of Gut Aiderbichl in Germany and Austria house well over 1,000 rescued animals, including a multitude of dogs that have found a new and safe place with us.

Allow me to apologize for some instances in this book when the line between man and dog becomes blurry. But bear in mind I'm writing about some very special relationships here. Each episode bears witness to all the similarities we share with animals, especially when it comes to love.

Die 10 Gebote
der Hundehaltung

von Michael Aufhauser

1. Du sollst dich informieren, ob Hundehaltung in deinem Haus erlaubt ist, Freilaufwiesen in der Nähe sind und die Nachbarn Hundegebell von Lärm unterscheiden können.

2. Du sollst dir darüber im Klaren sein, dass ein Hund mehr kostet als eine Schüssel Trockenfutter am Tag.

3. Du sollst beim Hundekauf nicht eitel und egoistisch sein, nicht nach Aussehen und Rasse entscheiden, sondern den Partner suchen, der dich 15 Jahre lang begleiten kann.

4. Du sollst deinen Hund nicht als Mittel zum Zweck verwenden: gegen Einsamkeit, Bewegungsmangel, als Kinderersatz oder Mediator bei Beziehungskrisen.

5. Du sollst deinen Teppichboden, dein makelloses Bettlaken, den Rasen im Garten und die Unversehrtheit deiner Autositze nicht über das Wohlbefinden deines Hundes stellen.

6. Du sollst dir Zeit nehmen für deinen Hund. Mindestens zwei Stunden täglich.

7. Du sollst deinen Urlaub auf deinen Hund abstimmen und wissen, dass Zusammenleben auch Verzicht bedeutet.

8. Du sollst deinem Hund Freundschaften zu anderen Hunden ermöglichen, auch wenn dir nicht jeder Hundebesitzer liegt.

9. Du sollst PFUI, SITZ, PLATZ und AUS als Notbefehle ansehen, nicht selber ständig den Alpha-Hund spielen, sondern Geduld und Einfühlung für die wichtigsten Mittel der Erziehung halten.

10. Du sollst deinen Hund nicht unterschätzen, sondern ihn als Persönlichkeit mit Gefühlen und Charakter achten. Ihn wenn irgend möglich nicht weggeben und für ihn gesorgt haben, solltest du vor ihm gehen müssen.

THE 10 COMMANDMENTS OF DOG OWNERSHIP

by Michael Aufhauser

1. You shall inquire if dog ownership is allowed at your residence, if off-leash areas are nearby, and if your neighbors are able to distinguish between barking and unwanted noise.

2. You shall be aware that owning a dog costs more than just a bowl of dry feed every day.

3. When buying a dog, you shall refrain from being conceited and self-centered and from basing your decision on looks and breed. Instead, look for a partner to be at your side for 15 years.

4. You shall not use your dog as a means to an end against loneliness, lack of exercise, as a substitute for kids, or as a mediator in rocky relationships.

5. You shall not value your carpet floor, your immaculate bed sheets, your lawn or the spotlessness of your car seats more than the well-being of your dog.

6. You shall dedicate time to your dog. At least two hours every day.

7. You shall arrange vacations to fit your dog, knowing that living together also entails certain sacrifices.

8. You shall allow your dog to befriend other dogs, even if you may not like all of their owners.

9. You shall think of NO, SIT, DOWN, and OUT as last resort commands and you shall not act as the perpetual alpha dog. Instead, focus on patience and empathy in training your dog.

10. You shall not undervalue your dog. Instead, respect it as an individual with feelings and personality, and should you be called home by the Lord before your dog is, ensure that it is in proper care instead of simply giving it away.

A Dog's Life— Abandoned in Spain

MÁLAGA – WIE ALLES BEGANN

Es war noch früh am Morgen und die Strände von Torremolinos an der Costa del Sol waren noch wie leergefegt. Ich hatte mich mit meiner Mitarbeiterin Yvonne verabredet, um die Ruhe zur Büroarbeit zu nutzen. Da bellte ganz laut ein Hund.

Wir liefen zum Fenster. Draußen hantierten fünf Männer in blauen Overalls mit einer Stange, an der oben eine Schlinge befestigt war. In der Schlinge zappelte ein Schäferhund, den sie eingefangen hatten, und schrie.

Die Männer warfen das Tier in einen Lieferwagen und fuhren langsam weiter, die Strandpromenade entlang. Offenbar hielten sie Ausschau nach freilaufenden Hunden und hatten den Auftrag, sie verschwinden zu lassen, mit welchen Mitteln auch immer.

Wir folgten dem Lieferwagen. Ich war damals, vor über 20 Jahren, noch Tourismus-Manager mit einer Dependance auch in Spanien. Ich war kein Tierschützer, aber wir kannten und mochten diesen Schäferhund. Wir hatten ihn immer mit Futter und Wasser versorgt, und außerdem war er am Strand der Liebling der Badegäste. Wir wollten ihn einfach wieder frei bekommen.

Die Fahrt ging Richtung Málaga. Am Ziel angekommen, fuhr der Lieferwagen durch ein großes Tor, auf dem „Perrera" stand. Die Anlage, die sich dahinter verbarg: mehrere Gebäude, die aussahen wie weiß gekalkte Baracken. Nachdem wir ins Innere gelangt waren, überließ uns der Chef des Ganzen den Hund gegen Bezahlung einer Impfgebühr, und Yvonne fuhr gleich mit ihm zurück.

Ich blieb. Mir kam die „Perrera", man kann es in Málaga schwer sagen, spanisch vor, aber jedenfalls dubios. Es gab viele Hunde. Die großen saßen alleine in Zwingern, die nach Desinfektionsmitteln rochen, die kleinen zu mehreren. Die Menschen dort waren mit ihren Aufgaben voll beschäftigt und nahmen mich kaum wahr. Mich interessierte ein ganz bestimmtes Gebäude, abgetrennt durch ein Tor, das gerade offen stand.

Da sah ich zwei Männer mit Gasmasken. Sie zogen tote Hundekörper aus einem großen Ofen. Ende einer Vergasung. Das also war das Geheimnis der „Perrera". Hier schickte

Málaga—How It All Began

It was still early in the morning and the beaches of Torremolinos at Costa del Sol were still deserted. I'd arranged to meet Yvonne, my coworker, to use the quiet to wrap up some office work. That was when a dog began to wail.

We ran to the window. Outside, five guys in blue overalls were handling a rod with a loop at one end. Squirming in that loop was a German Shepherd they had caught, and it was just wailing.

The men threw the dog into a van and started driving slowly along the beach. Apparently they were on the lookout for stray dogs and their job was to get rid of them by any means necessary.

We decided to follow the van. This was more than 20 years ago when I was a tourism manager with a branch office in Spain. Although my agenda didn't include animal rights, we knew and liked that German Shepherd. We'd always hooked him up with food and water, and besides, he was a favorite among the beach goers. We just had to find some way to free him.

The van headed to Málaga. Reaching its destination, the van pulled through a large gate marked "Perrera." The property concealed behind it consisted of a number of buildings looking like white limestone barracks. Since we'd made our way inside, the man in charge of the whole outfit let us have the dog in return for a vaccination fee and Yvonne immediately took the dog back to our place.

I stayed. Something about this "Perrera" place just didn't feel right. It had a lot of dogs. The big ones sat by themselves in kennels reeking of disinfectants while the small ones had to share theirs. The workers there were fully absorbed in their jobs and barely noticed me. One building in particular raised my curiosity even more. It was set apart from the rest by a gate of its own, which happened to be open.

There I saw two men with gas masks. They were pulling the bodies of dead dogs from a large furnace. The final outcome of a gas chamber. This was the secret behind "Perrera."

man Hunde in die Gaskammer. Als die Arbeiter sahen, dass sie beobachtet wurden, schlossen sie das Tor.

So etwas hatte ich noch nie gesehen und hätte auch nie gedacht, dass so etwas geschieht. Eine Tierärztin der „Perrera" erklärte mir dann, streunenden Hunden, die hier eingeliefert werden, bliebe eine Frist von drei Tagen. Holt sie keiner ab, werden sie getötet. Sie zeigte mir die Baracke, in der die nächsten Todeskandidaten warteten.

It was a place for sending dogs into a gas chamber. Realizing they were being watched, the workers shut the gate.

I had never seen anything like it and I never imagined anything like this could be possible. Then a veterinarian at the "Perrera" told me that stray dogs that are brought here get a grace period of three days. If nobody claims them, they are killed. She also showed me the barrack where the next candidates awaited their deaths.

I looked into the kennels. A female Galgo showed me the stump where one of her legs had been torn off, revealing raw bone. For some of the others, I was positive I'd find some good homes. That poor female Galgo, however, really might be better off dead, I thought, given my inexperience at the time and the shock I felt at the sinister goings-on at the facility. Of course, I'd never entertain such a thought today, but what was I to do back then? I spontaneously offered the veterinarian to buy all the healthy dogs. She sold them to me by charging me vaccination fees; in other words, $ 30 for each dog. I paid for a total of 40 animals.

I then took a taxi to the airport, rented a cargo van and we loaded the dogs in it shortly before "Perrera" closed for the night. Frightened, they all sat in the van without making a single sound. But just as we made our exit through the large gate, their tension began to subside. Then a female Galgo and a poodle mix hopped on the driver's seat with me and began licking at me. I pulled over into the shade of a large tree and did my best to comfort the dogs.

How could they be sure they were safe? How could any dog feel comfortable being transported with other dogs by some stranger in a cargo van? But somehow they just did —maybe it was my breath, maybe it was my gestures or maybe they heard a different tone of voice. Good Lord!—I thought—what are we doing to living beings with such sensitive senses? And what kind of person would demand these acts from the employees of a municipal facility?

That experience changed my life. My image of the world had been shattered. I set my life in a new direction.

Ich schaute in die Zwinger. Eine Galgo-Hündin streckte mir den Stumpf ihres abgerissenen Beines entgegen. Man sah den blanken Knochen. Bei anderen war ich mir sicher, dass ich gute Plätze für sie finden würde. Für die arme Galgo-Hündin könnte der Tod eine Erlösung sein, dachte ich mir, unerfahren, wie ich war und schockiert von den finsteren Vorgängen in der Anlage. So ein Gedanke würde mir heute nicht mehr kommen – aber was konnte ich damals denn tun? Spontan bot ich der Tierärztin an, die gesunden Hunde zu kaufen. Ich bekam sie alle um den Preis einer Impfgebühr, also für 20 Euro pro Hund. Ich bezahlte insgesamt für 40 Tiere.

Dann nahm ich ein Taxi zum Flughafen, mietete einen Lieferwagen und kurz vor Dienstschluss in der „Perrera" luden wir die Hunde ein. Eingeschüchtert saßen sie im Wagen und machten keinen Mucks. Aber als wir durch das große Tor nach draußen fuhren, löste sich die Anspannung. Eine Galgo-Hündin und ein Pudel-Mix sprangen zu meinem Fahrersitz und leckten mich ab. Ich stoppte im Schatten eines großen Baumes und versuchte, die Hunde zu beruhigen.

Woher wussten sie, dass sie außer Gefahr sind? Es hat doch für einen Hund nicht automatisch etwas Vertrauenerweckendes, wenn er mit anderen zusammen von einem Fremden in einem Lieferwagen befördert wird. Sie müssen es einfach gespürt haben, vielleicht an meinem Atem, durch meine Gesten, einen anderen Ton des Umgangs. Um Himmels willen, dachte ich, was tun wir Lebewesen an, die mit solchen Sinnen ausgestattet sind. Und was sind das für Menschen, die von den Angestellten dieser städtischen Anlage solche Arbeiten verlangen?

Nach dieser Erfahrung änderte sich mein Leben. In mein Bild von der Welt hatte der Blitz eingeschlagen. Ich setzte meinem Dasein andere Ziele.

Die Hunde verteilte ich auf Hundepensionen und Heime. Später brachte ich sie nach Deutschland, wo ich sie gemeinsam mit Tierschützern an gute Plätze vermittelte. Zwei Hunde dieser ersten Rettung blieben bei mir: ein renitenter Pudel-Mix namens Rodrigo und die Galgo-Hündin Prinzi.

Ich kehrte dann noch oft nach Málaga zurück, bemüht um Intervention, und rettete noch viele Hunde. Denn die Gaskammer arbeitet weiter.

I placed the dogs in various dog sanctuaries and shelters. Thereafter I brought them to Germany where I worked with animal rights activists to find good homes for them. Two of those dogs from that first rescue stayed with me: Rodrigo, a rebellious poodle mix, and Prinzi, the female Galgo.

After that, I made a lot of trips back to Málaga. I was intent on intervention, rescuing scores of other dogs. That gas chamber was still operating, after all.

Prinzi, das Leid der Windhunde

Manche Tiere leiden unter den typisierenden Blicken der Menschen, die alles schubladisieren. Dazu gehören auch die Windhunde. Dort, wo sie zu unserem angeblichen Nutzen eingesetzt werden, dienen sie als „Rennmaschinen": Die Greyhounds auf Hunderennbahnen und die Galgos als Jagdhunde, hauptsächlich in Spanien. Dass diese Hunde, genauso wie alle anderen, Individuen sind, wollen wir entweder nicht wahrhaben oder es hat niemand Zeit, darüber nachzudenken.

Ihr Leid beginnt aber erst nach ihrer eigentlichen „Karriere" richtig, wenn sie nicht mehr gebraucht werden. So haben Greyhounds durch ihre gute Kondition ein großes Herz, was sie für Versuchslabore interessant macht. Und weil sie auch sonst selten Fürsprecher haben, setzt sich auch dann kaum jemand für sie ein.

Galgo-Jagdhunden, die nicht mehr flink genug, verletzt oder schon alt sind, werden nicht selten von ihren Besitzern Schlingen um den Hals gelegt. Man hängt sie an den nächs-

ten Baum, wo sie sich zu Tode strampeln. Oft tagelang.

Prinzi saß hinter Gittern in der „Perrera" von Málaga, als wir uns trafen. Wie sich später herausstellte, litt sie an Arthrose und ist wahrscheinlich deshalb zum Vergasen dorthin gebracht worden.

Nach ihrer Befreiung, auf dem Weg zu einer Hundepension in einem Transporter, den ich selbst fuhr, kam sie immer wieder zu mir und blickte mich an.

Prinzi and the Plight of the Greyhounds

Some animals suffer just as well under the stereotyping of people who need to categorize the whole world. Among these animals are the Greyhounds. In some places, where they supposedly serve the benefit of mankind, they end up used as "racing machines": The Greyhounds are used on dog racing tracks and the Galgos are used as hunting dogs, primarily in Spain. The truth is that these dogs are individuals just like all the others. The problem is we either refuse to acknowledge it or we're all just too busy to give it any thought.

Actually, the real suffering doesn't begin for these dogs until after their "careers"—when they're no longer needed. Greyhounds, for instance, have large hearts due to their athletic conditioning, putting them in high demand for test labs. And since few people ever bother to speak out on their behalf anyway, hardly anyone is willing to address the problem.

It is not rare for Galgo hunting dogs that have lost their agility, were injured or are just old to end up with nooses put on their necks by their owners. They are then hung from the nearest tree where they slowly writhe to death—in many cases, for days.

Prinzi was behind bars in Málaga's "Perrera" when I met her. As it turned out, she suffered from arthrosis, making it very likely that she was there to be gassed.

After being freed and on her way to a retirement home for dogs in a van driven by yours truly, she constantly approached me and looked at me. As utterly inexperienced as I was back then, I still found myself unable to resist her gazes. It was like she was simply yearning to be acknowledged. It must have been her past. Even then it dawned on me that animals, like children, are not in conflict with their thoughts, feelings and their instinct. She was able to reach out and touch my heart.

In those days, taking a dog into any relatively clean hotel in southern Europe didn't come without its challenges, unless you had a pure breed and it was your private dog. Prinzi, on the other hand, was mere skin and bone, she had bruises on her and she looked nothing like a private dog. You are what you wear, I thought to myself, so I drove to some

Ich war damals völlig unerfahren und dennoch konnte ich ihren Blicken nicht widerstehen. Sie hatte so etwas wie eine Sehnsucht nach Wahrnehmung. Das muss mit ihrer Vorgeschichte zu tun haben. Schon damals dämmerte mir, dass Tiere, wie auch Kinder, keine Konflikte haben mit ihren Gedanken, Gefühlen und ihrem Instinkt. Sie konnte sich verständlich machen und erreichte mein Herz.

Damals war es so, dass man im Süden Europas nicht ohne weiteres einen Hund in ein relativ ordentliches Hotel mitnehmen konnte. Es sei denn, es handelte sich um einen Rassehund, den man von daheim mitgebracht hatte. Doch Prinzi war abgemagert, hatte Schürfspuren und war ganz offensichtlich kein Privathund. Kleider machen Leute, dachte ich, und fuhr nach Marbella zu einer dekadenten Hundeboutique. Dort gab es das Ego-Halsband für Hundebesitzer, mit funkelnden Glassteinen besetzt. Für die ersten Tage allerdings war ich dankbar, so ein Accessoire aufgetrieben zu haben. Prinzi schien den Plan zu verstehen, und so schritten wir erhobenen Hauptes durch die Marmorhalle des Hotels. Die Groteske kam an, und der Concierge nickte freundlich, als er uns beide kommen sah.

Ich lebte aus beruflichen Gründen in diesem Hotel und stand kurz vor der Abreise in die Schweiz, wo ich die meiste Zeit des nächsten halben Jahres in Gstaad verbrachte. Prinzi begleitete mich dorthin. Ich war dort nicht großartig abgelenkt und konnte auf das Wesen der Galgo-Hündin eingehen.

Sicherlich haben Tiere genauso wie Menschen Potentiale, mit denen sie geboren werden, ganz persönliche. Prinzi war anständig, sanftmütig, edel und fair. Wie konnte diese Hündin jemals damit umgehen, dass sie von ihren Vorbesitzern als ganz „normaler" Jagdhund eingestuft wurde? Ich kann mir vorstellen, dass sie, abgesehen von physischem Leid, dem sie ausgesetzt war, auch darunter leiden musste.

Prinzi brachte mir bei, dass es gar nicht so schwierig ist, die Welt aus der Sicht eines Hundes wahrzunehmen und bei der Beurteilung eines Lebewesens nicht nur den eigenen Vorstellungen zu folgen. Es war für mich eine ganz neue Art zu denken, zu der mir auch Prinzi, die Galgo-Hündin, verhalf.

tacky dog boutique in Marbella. There I found what could only be called the "ego collar for dog owners", judging by the sparkling glass stones on it. Nonetheless, I was glad I had this little accessory—for the first couple of days, anyway. Prinzi seemed to understand my plan and we paraded through the marble lobby of the hotel, with our heads held high. Our little comedy act seemed to work, because the receptionist nodded friendly when he saw the two of us approach.

I stayed at that hotel for business reasons before leaving for Switzerland, where I spent the better part of the next six months in Gstaad. Prinzi accompanied me there. Without any major distractions that might appeal to me, I found the time to study the character of this female Galgo.

I am convinced that animals come into this world with the same kind of potential that people have—indeed, highly individual potential. Prinzi was well-behaved, gentle, noble, and fair. How on earth did she ever put up with previous owners, who lazily thought of her as "just

another" hunting dog? I can imagine how that must have fueled her misery, not to mention the physical pain she must have felt.

Prinzi taught me that it's not all that hard to see the world through a dog's eyes and to think outside of one's own standards when placing judgement on another living being. My eyes opened to a whole new perspective, and Prinzi, the female Galgo, played no small part in opening them for me.

„ALS ICH DEM LEID DER HUNDE IN MÁLAGA BEGEGNETE, DACHTE ICH: WIE UNMENSCHLICH UND ZYNISCH GEHT MAN EIGENTLICH MIT DEM LEBEN UM. DA SCHLUG EIN BLITZ IN MEIN LEBEN EIN."

"WITNESSING THE ORDEAL OF THOSE DOGS IN MÁLAGA, I COULDN'T HELP BUT THINK ABOUT HOW INHUMANE AND CYNICAL WE ARE IN DEALING WITH A LIFE. IT ALTERED MY LIFE IN A VERY PROFOUND WAY."

Rodrigo, die Mitte zwischen Sir und Clochard

Zu den ersten Hunden, die ich vor einer Gaskammer rettete, gehörte ein Pudel-Mix, der wie ein Schaf aussah und am Bauch schwarz gepunktet war. Ich brachte ihn zu einer netten Dame in ein Tierheim in Spanien. Dort sollte er bleiben, bis ich für ihn einen Flugtransfer nach Deutschland fand.

Tags drauf besuchte ich ihn und die anderen Hunde, die dort untergekommen waren. Die Tierheimleiterin lobte die Hunde und dass sie sich in ihren Gruppen schon eingewöhnt hätten – bis auf einen. Das Schäfchen, das jetzt Rodrigo hieß. „Bei aller Hilfsbereitschaft, der kann leider nicht hier bleiben."

Was war geschehen? Er kann ja nicht sprechen, aber er hat uns durch sein Verhalten etwas sehr Wichtiges mitgeteilt: Ich gehöre nicht in einen Zwinger. Dem war er nämlich sportlich entkommen, schmuggelte sich in die Tierklinik des Tierheims, nahm sich dort jede Menge Handtücher aus dem Regal und baute sich daraus ein weiches, großzügiges Körbchen. Als man ihn dann wieder zurück in den Zwinger brachte, bellte und schrie er so unerträglich, dass er in der Folge mein persönlicher Hund wurde.

15 Jahre sollte er bei mir bleiben. Einmal, erinnere ich mich, war er mir an der Costa del Sol weggelaufen. Er wurde über Radio und Plakate gesucht. Nach zwei Tagen entdeckte ich ihn endlich. Stumpfsinnig lief ich, immer seinen Namen „Rodi, Rodi" rufend den Strand entlang, als er völlig verdreckt aus einem riesigen Kanalrohr kam. Gemeinsam mit einer Gang von Rüden verfolgte er eine läufige Hündin.

Seine Wiedersehensfreude hielt sich in Grenzen, und ich hatte große Mühe, zu verhindern, dass er mit den anderen Hunden weiterzog. Das ist nur eine von vielen Episoden, die ich über Rodrigo erzählen könnte, der genau die Mitte zwischen Sir und Clochard war.

Rodrigo, midway between Sir and Clochard

One of the first dogs I rescued from the gas chamber was a poodle mix that looked like a sheep with a pattern of black spots on its belly. I took him to a nice lady, who ran an animal shelter in Spain. My idea was to keep him there until I found a flight that would take him to Germany.

The next day, I looked in on him and the other dogs in the shelter. The owner of the animal shelter praised the dogs and the fact that they all fit in well with their respective groups—all but one. My little buddy, the "sheep"—now called "Rodrigo." "As much as I'd love to help you, he can't stay here, I'm afraid."

What had happened? Well, since he can't talk, he used his behavior to tell us something of grave importance: I'm not cut out to be in a kennel. And it was a kennel he had swiftly escaped from. Then he had sneaked into the hospital ward of the animal shelter, snatched as many towels as he could and piled them into a large, soft basket. On being returned to his cage, he barked and howled so wretchedly that he became my personal dog.

He lived by my side for 15 years. I remember one time when he ran away from me at Costa del Sol. A search for him involved flyers and radio stations. Two days went by before I finally located him. There I was, doggedly running down the beach hollering "Rodi, Rodi"—his name—when he came out of a huge drainage pipe, dirty all over. Accompanied by a pack of males, he'd been on the pursuit of a female in heat.

He wasn't exactly thrilled to see me again and I had a tough time to keep him from joining those other dogs. This is just one of many episodes I could tell you about Rodrigo, who was exactly midway between Sir and Clochard.

BURSCHI, MEIN SCHRECKLICHER LIEBLINGSHUND

Zum Leben gehören gute Vorsätze, und so nehme ich mir immer wieder vor, keines der 1000 Tiere, die auf den Aiderbichler Gütern unter meinem Schutz stehen, zu bevorzugen. Es gibt kein Lieblingstier, sage ich mir, es gibt nur Lieblingstiere. Jedes Lebewesen ist auf seine eigene Weise mein Favorit, und ich muss schon selbst staunen, wie erfolgreich ich meinem Vorsatz treu bleibe.

Ein einziges Mal gab es eine Ausnahme: Burschi, ein Terrier-Mix, nicht gerade der Freundlichste unter der Sonne. Ganz im Gegenteil: Er war verschlagen, hartnäckig und schreckte nicht vor unfeinen Methoden zurück, um auf sich aufmerksam zu machen. Burschi war einer von weit über 1000 Hunden, die ich vor dem Tod in den Gaskammern Málagas bewahrte.

Im Belly meiner Linienmaschine von Spanien in die Schweiz befanden sich bereits drei große Jetboxen mit geretteten Hunden. Da kam einer meiner Mitarbeiter mit Burschi an der Leine und bat mich, ihn mitzunehmen. Er sei so winzig und arm, und außerdem dürfe er mit in die Kabine. Also saß er wenig später, traurig und mit hängendem Kopf, erst in einer Tasche, dann unerlaubter Weise auf meinem Schoß: Burschi, auf der Reise in seine neue Welt.

In der Schweiz nahm ich ihn mit nach Hause zu den anderen Hunden, die aus irgendwelchen Gründen schwer vermittelbar waren. Burschi aber sah so entzückend aus, dass ihn gleich tags darauf ein bezauberndes Ehepaar mit genügend Zeit und einem Garten zu sich nahm. Ich war glücklich, hatte aber die Rechnung ohne Burschi gemacht.

Bald schon stand er schwanzwedelnd wieder mit dem Ehepaar vor meiner Tür. Die älteren Herrschaften waren zwar vollkommen tierlieb, sahen sich aber überfordert, weil Burschi zwar kaum gefressen, aber nahezu jeden gebissen hatte. Er selbst war bei bester Laune, sprang auf mein Bett und zeigte, dass ein kleiner, selbstbewusster Bursche wie er haargenau weiß, wo er zu Hause ist. Nämlich bei mir.

BURSCHI, MY TERRIBLE FAVORITE

We all need positive resolutions in life, and one of mine is never to pick favorites among any of the 1,000 animals under my care at Gut Aiderbichl. I tell myself there's no such thing as one favorite animal, that all animals are favorites. Every creature is a favorite of mine in its own way, and sometimes I'm pretty amazed at my own ability to abide by my resolution.

The one and only time that ability failed was when Burschi, a terrier mix, entered my life. That's not to say Burschi was a regular sweetheart—far from it. He was sly, stubborn and didn't shy away from improper means of seeking attention. Burschi was one of well more than 1,000 dogs I'd rescued from death in the Málaga gas chambers.

The belly of my scheduled aircraft taking me from Spain to Switzerland already contained three large jet boxes full of rescued dogs. Then one of my colleagues came along with Burschi on a leash and asked me to take him with me, arguing how small and pitiful he was, and besides, he was allowed in the cabin. That's how this dog ended up with me shortly afterward—his head lowered and looking sadly out of a duffel bag before finally settling on my lap, without permission. That was Burschi on his way to his new world.

In Switzerland, I brought him home to the other dogs, which, for some reason or other, had a tough time finding new homes. Burschi, on the other hand, looked so adorable that a lovely couple with plenty of time to spare and with a garden adopted him the very next day. I was relieved, not knowing that Burschi had other plans.

Nach zwei weiteren Vermittlungsversuchen gab ich auf. Von nun an lebte Burschi in meinem Privathaus. Man darf aber nicht glauben, dass ihn die Erfüllung seines Wunsches handsam gemacht hätte. Er hatte seine Erfahrungen und verhielt sich entsprechend.

Er war äußerst dominant und ging sogar Schäferhunde an. Berner Sennenhunde zogen ihren Schwanz ein, wenn der Zwerg mit kreischendem Gebell auf sie zukam. Er sah mich dann stolzen Blicks an und wollte gelobt werden. Ich aber schimpfte mit ihm, weil ich doch wollte, dass er sich zu einem sozialen Wesen entwickelt, aber Burschi blieb hart.

Er änderte allerdings seine Taktik, weil er begriffen hatte, dass die Opferrolle besser zieht und setzte infolgedessen auf Mitleid. Als Ersten traf es Rodrigo, einen besonders lieben Pudel-Mix, ebenfalls vor der Gaskammer gerettet. Er war zwar intelligent, aber von beispielloser Gutmütigkeit. Einmal schrie Burschi auf, winselte, weinte und obendrein humpelte er auch noch. Noch lauter schrie er, als ich versuchte, ihn wie einen Schwerverletzten auf den Arm zu nehmen. In der Ecke im Körbchen lag Rodrigo mit gesenktem Kopf, als würde er mir seine Schuld gestehen und um Verzeihung bitten. Natürlich wies ich ihn zurecht und ließ ihn zur Strafe ein paar Stunden links liegen.

It wasn't long before the couple was back at my door with a tail-wagging Burschi. As much as this elderly couple loved animals, they were at their wit's end, because he'd barely touched any food they gave him, having taken to biting people instead. Burschi himself, looking like a million bucks, hopped onto my bed to

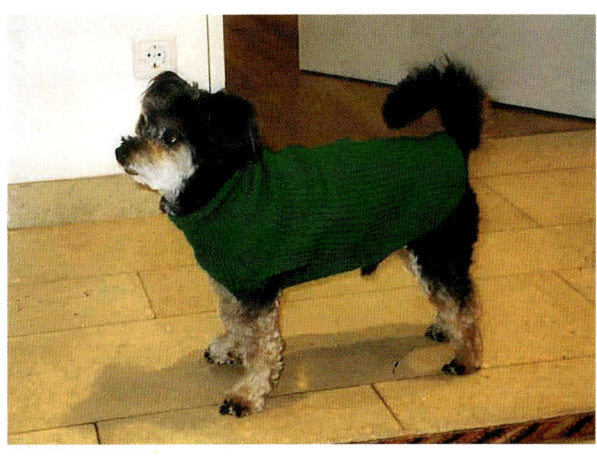

show me that a little, self-confident dude like himself knew perfectly well where his home was—in *my* house.

When another two attempts to place him proved futile as well, I surrendered. From now on, Burschi would be living in my private home. However, if you think seeing his wish come true made him any easier to handle, think again: He knew the ropes and acted accordingly.

Burschi was extremely dominant, even taking on German Shepherds. Bernese Mountain Dogs lowered their tails when this little guy met them head on with his screeching barks. Then he turned to me with that proud look on his face and expected me to praise him. Instead, I came down hard on him, because, after all, I was trying to teach him to be social. Burschi, however, was having none of it.

Rather, he changed his tactics after realizing it was better to play the role of the victim and to count on the pity that was likely to follow. His first victim was Rodrigo, a particularly friendly poodle mix, who'd been rescued from the gas chamber just like Burschi. Although Rodrigo was smart, he was also exceptionally softhearted. In one case, Burschi was howling, whimpering, crying, and on top of it all, even limping. He howled even louder when I tried to hold him in my arm as if he'd been gravely wounded. Lying in his basket in the corner was Rodrigo, his head lowered as if making a confession and asking for mercy. Not knowing any better, I gave him an earful and punished him by giving him the cold shoulder for a few hours.

Burschi galt von nun an als der arme Prügelknabe, forderte dafür Respekt und erhielt Sonderbehandlung. Verließ ich das Haus ohne Hunde, kam Burschi mit. Er musste doch beschützt werden, durfte als einziger Hund in meinem Bett schlafen, wenn er wollte, und mich sogar auf großen Reisen begleiten. Bis, ja bis ich ihn per Zufall dabei beobachtete, wie er einen anderen Hund ins Bein zwickte und dann selber laut aufschrie. Oh, Burschi! Oh, diese Überlebenstechniken eines Underdogs!

An meinen Gefühlen für ihn änderte sich durch die Entdeckung seiner Tricks nichts. Ja, Burschi war mein schrecklicher Lieblingshund. Das musste ich mir eingestehen, aber zugleich allen anderen zum Ausgleich mehr Zuneigung schenken, als es mir bis dahin möglich schien. Ich war es ihnen und mir schuldig, und sie nahmen meine ehrlichen Bemühungen wahr.

Mit Burschi war das komplizierter. Er war vollkommen unbestechlich. Auch meine Mitarbeiter, die ihn fütterten und sich rührend um ihn bemühten, galten ihm nichts, sobald ich da war. Aber auch ich konnte in Ungnade fallen. Burschi hasste z.B. Sentimentalitäten.

Burschi had now established himself as the poor whipping boy. As such, he demanded respect and was treated special. If I left the house without dogs, Burschi was right there. After all, wasn't he the one that needed protection, wasn't he the only dog permitted to sleep in my bed whenever he felt like it and wasn't he even permitted to come with me on major trips? Indeed, he was—until I saw him nip another dog in the leg, only to howl out himself. Oh, Burschi! Talk about the survival techniques of an underdog!

Seeing through his tricks never changed my feelings for him. Yes, Burschi was my terrible favorite. I had to admit it to myself and I had to make it up to all the other dogs by showering them with a kind of affection I didn't even know I had in me. But that was what I owed them as well as myself and they didn't fail to acknowledge my sincere efforts.

Burschi, himself, was more complex. He was completely resolute. Any effort by my staff to feed him and shower him with affection meant nothing to him once I was there. But even I could lose favor with him. For one thing, Burschi was loath to sentimentalities. Once, when he was genuinely sick and lying in his basket in a weakened state, I laid down on the floor

Als er einmal richtig krank war und schwach in seinem Körbchen lag, legte ich mich neben ihn auf den Boden. Ich bin für solche Theatralik eigentlich nicht gebaut, erinnerte mich aber an meine Mutter und daran, wie sie sich verhielt, wenn es mir in Kindertagen schlecht ging. Obwohl als Sänger nicht begabt, setzte ich trotzdem zu einem Schlaflied an und streichelte Burschi. Da kehrte das Leben in ihn zurück, er sprang mit feurigen Augen hoch und biss mir in die Nase. Vielleicht war Burschi musikalisch.

Einmal verteidigte er mich tapfer und raubeinig gegen Einbrecher, wirkte aber ziemlich verstimmt, weil er mich schon vorher gewarnt hatte, worauf ich nicht reagiert und ihn zu beruhigen versucht habe. Burschi war schlau. Manchmal sogar schlauer. Der Einbruch verlief dann durch seinen Einsatz glimpflich.

next to him. Although I'm usually not the type of person for such theatrics, I remembered what my mother used to do whenever I came down sick as a child. And not being much of a singer either, I still tried to sing a lullaby while I was stroking Burschi. That was when he suddenly came alive again, as he jumped up with fiery eyes and bit me in the nose. Maybe Burschi had an ear for music.

He even put up a brave and fierce struggle to defend me from some burglars once. Afterward he appeared rather disgruntled, because he had tried to warn me earlier and I had failed to react and tried to calm him instead. Burschi was clever. Sometimes beyond clever. It was his action that kept the burglary from turning into something more serious.

Burschi lived to be around 19 years old. During all those years I spent with him, he had taught me a lot. Like never giving in without a solid reason. Like staying loyal to my goals and my ways.

Burschi wurde etwa 19 Jahre alt. In der langen Zeit mit ihm hatte ich viel gelernt, zum Beispiel, nicht ohne guten Grund nachgiebig zu sein, an Zielen festzuhalten und meinem Weg treu zu bleiben.

Am Ende musste ich ihm zuliebe dann doch noch den Chef spielen. Das war, als der Tag des Abschieds kam, und ich wusste, dass er nicht mehr konnte. Wir machten uns gemeinsam auf seinen letzten Weg. Ich hatte alles bestens arrangiert. Er sollte nichts bemerken.

Die Tierärztin traf uns auf einer Wiese, weit weg von ihrer Praxis. Müde sah Burschi sie an und dachte vermutlich: „die schon wieder!". Ich hielt ihn in meinen Armen, was er womöglich lächerlich fand, und die Ärztin gab ihm sein Lieblingsgutti. Er hatte kaum noch die Kraft, um es anzunehmen. Burschi bemerkte nichts, als die Ärztin ihm ein Schlafmittel gab. Dann fuhren wir beide ein letztes Mal gemeinsam mit dem Jeep durch die Landschaft, bis er ganz fest schlief.

Er hat nicht mitbekommen, dass wir schließlich in der Praxis ankamen und er dort erlöst wurde. Alles geschah ganz friedlich. Diesmal war ich schlauer als er. Aber nur vielleicht, denn als er vor mir lag, sagte mir sein Gesichtchen: Hast du gut gemacht, Papi! Hast viel gelernt, und übrigens: Danke!

Immer wenn ich Burschis Geschichte erzähle, denke ich mir: Von wem ist da eigentlich die Rede? Von einem Vierbeiner? Ich glaube, wenn wir über Tiere reden, dann erzählen wir in gewisser Weise weiter, was uns die Tiere selbst mitgeben. Sie sind Ratgeber, die uns viel über uns selbst und das Leben im Allgemeinen beibringen. So wie Burschi.

As it turned out, I did have to show him who's in charge—for his sake. It was the day when the time had come to say farewell and I knew he'd reached his end. Together, we were ready for his last journey. I had arranged everything perfectly. He wouldn't notice a thing.

The veterinarian met us on a pasture far away from her practice. Burschi gave her a tired look, probably thinking, "What—*her* again?!" I held him in my arms, him probably thinking what a tired act *that* was, when the vet gave him his favorite snack. He barely had enough strength left to accept it. Burschi never noticed the sedative she gave him. Then I took him on his final ride through the countryside in my Jeep before he fell into a very deep sleep.

He didn't notice how we eventually pulled up to the veterinarian's practice where he was put out of his misery. It all went very peacefully. This time, I outwitted *him*. Then again, maybe I didn't, because as he lay there before me, the look on his adorable face said, "Nice work, pops! You sure learned a lot! And by the way, thank you!"

Whenever I tell Burschi's story, I can't help but wonder: whose story am I really telling. That of a canine? I believe that when we talk about animals, all we really do is to pass on what the animals pass on to *us*. They're like mentors who not only teach us a lot about ourselves but also about life in general, the way Burschi did.

MEINE MYSTISCHEN HUNDE

MY MYSTICAL DOGS

FLOPPY, EINE HÜNDIN, DIE IHR LEBEN SELBST BESTIMMTE

In meinem Stadthaus lebte Floppy, Elfriede Jelineks Hund. Die Hündin der österreichischen Nobelpreisträgerin, ein Bearded Colly, kam zu uns, weil sie ein Leben mit nur einem einzigen Menschen nicht führen konnte. Die souveräne Neigung Elfriede Jelineks zum Separaten konnte die Hündin nicht mitleben. Sie wollte andere Hunde kennen lernen und einen anderen Tagesablauf, und so fiel dann zugunsten Floppys schweren Herzens die Entscheidung, sie zu mir nach Salzburg zu bringen.

Floppy wusste, was sie wollte. Sie war für Kompromisse nicht gebaut, suchte Kontakte, aber nicht das Rudel. Sie war, sagen wir, eine eigenschöne Persönlichkeit, distanziert sozial. Da hätte kein Salzburg-Besucher kommen können und fragen: Darf ich dich streicheln? Ihr Weltbild war weit entfernt von Sentimentalitäten. Eher von Widersprüchen geprägt.

Ein ehemaliger Dauergast in meinem Nebenhaus sagte mir, dass es nicht leicht sei, mit Floppy zu spielen. Sie bringt den Ball, hält ihn zwischen den Zähnen und gibt ihn nicht her, es sei denn, man erklärt sich zur Wurfmaschine auf Stunden. Das war ihr Spiel.

FLOPPY, A DOG IN CHARGE OF HER OWN LIFE

Floppy used to live in my town residence and her mistress was none other than Elfriede Jelinek, the Austrian Nobel Prize winner. Her pet, a Bearded Collie, came to us because she couldn't share her life with just one person. This female dog simply wasn't cut out for Elfriede Jelinek's indomitable thirst for individuality. Floppy yearned to meet other dogs and to live a different daily routine. So the heavy-hearted decision was made for Floppy's sake that she be brought to my place in Salzburg.

Floppy knew what she wanted. Never the type to compromise, she was out to make contact, although not with packs. Her personality was characterized by, shall we say, self-awareness and distinct aloofness. Any visitor to Salzburg trying to pat her would have been sheer out of luck. Floppy's world was far removed from sentimentalities. If anything, it was a world of ambivalence.

A former long-term neighbor of mine once told me that playing with Floppy was not an easy task. She'd return fetched balls by clenching them between her teeth and not let go of them without him acting as her ball projector for hours on end. That was her game, even if the ball didn't come into play. Even so, she still got to taste what hardly any living creature can do without—moments of affection. I can attest to that.

Thunder and lightning put her in an indescribable state of panic. As frightening as these highly charged spectacles may seem to all dogs, it was like sheer torture to Floppy. Her

Im Grunde ein Spiel ohne Ball. Trotzdem hat sie, was sich ein Lebewesen schwer entsagen kann, Momente der Zuneigung genossen. Ich bin Zeuge.

Wenn es donnerte und blitzte, geriet sie in unsägliche Panik. Dieses Feuerwerk, ein Schrecken für jeden Hund, war ihr das Höchste an Qual. Ihre Ängste steigerten sich mit dem Fortschreiten ihrer Wahrnehmung. Schon wenn eine Autotür zugeschlagen wurde, ging es ihr nicht gut. Aber sie konnte auch vergessen und fröhlich sein.

Sie zog dann, weil wir vermuteten, es wäre ihr Wunsch, in die zum Haus gehörende Wohnung meines Hundepflegers Hans Eder. Das Undenkbare geschah. Floppy, distanziert und eigenwillig, schlief bei ihm auf seinem Bett. Nur so fand sie Ruhe in der Nacht. Tagsüber genoss sie die Ausflüge auf die Freilaufwiesen und tobte sich aus.

Dann fand in einer nahegelegenen Kirche eine Hochzeit statt, mit allem Drum und Dran und Böllerschüssen. Floppy sprang in Panik über den Zaun und wurde überfahren.

Es hat schon oftmals geböllert, und dann waren es Kanonen. Floppy hat vieles zu Ende gedacht. Es war ihr Ende. Zu früh, aber nach einem intelligenten und selbstbestimmten Leben.

anxiety had a way of rising in tandem with her awareness. Even a car door being slammed put a knot in her stomach. Still, there was an easy-going and joyous side to her too.

Thinking it was what she wanted, we put Floppy in the care of Hans Eder, my dog trainer, whose apartment was part of my residence. What happened next was unthinkable. Floppy, aloof and indocile, went to sleep right on his bed. It was the only way she could rest at night. In the daytime, she loved going to off-leash areas and having a romp.

Then a wedding took place at a nearby church one day. It featured all the usual amenities along with loud booming sounds. In a state of panic, Floppy hopped a fence and ended up being run over.

While plenty of events involve fireworks, this one involved cannons. Floppy thought a lot of things through to the end. This proved to be her end. Her life may have been cut short, but not before she lived it with intelligence and self-determination.

Pirosch, der Jagdhund,
der die Hasen leben liess

Einer der ersten Hunde in meinem Leben war Pirosch, ein ungarischer Vorstehhund. Das war lange bevor ich mit dem Leid der Hunde in Berührung kam und deshalb war er auch mein einziger Hund, den ich aus einer Hundezucht holte. Seine Züchterin hieß Lilly Wossala. Um Farbe, Blesse und Abstammung zu prüfen, wollte ich Pirosch persönlich bei ihr in Budapest abholen. Als ich bei ihr eintraf, führte mich Lilly in einen Raum voll mit Campingliegen, und sie erklärte mir, dass Viszla, wie die Rasse auf Ungarisch heißt, in Körbchen, Zwingern oder auf Decken am Boden nicht glücklich werden.

Das ist jetzt über 30 Jahre her, und ich muss selbstkritisch eingestehen, dass mich damals manchmal der Spießer übermannte. Aber dann landete der Zufall gleich einen Volltreffer. Pirosch wollte gar nicht jagen. Im Gegenteil, zu seinem besten Freund wurde mein Kaninchen Hansi. Obendrein hatte er Persönlichkeit genug, sein Manko ganz selbstverständlich zu behaupten. Dank Pirosch, mit dem ich 16 Jahre meines Lebens verbringen

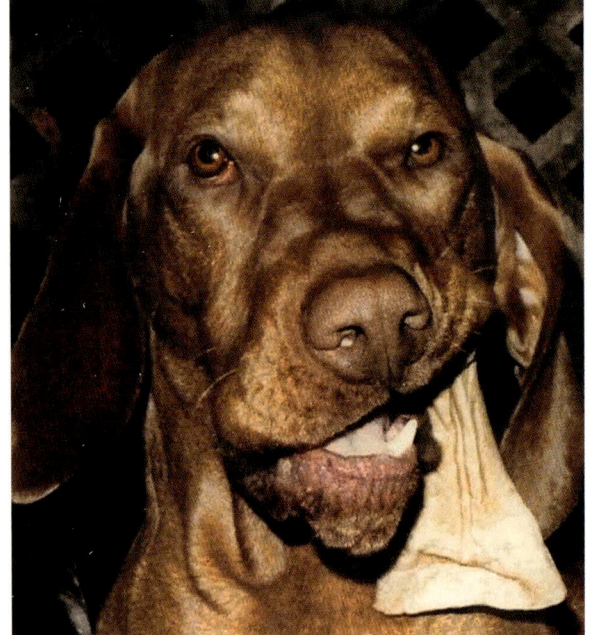

durfte, wusste ich schon bald: Hunde sind Individualisten, wie wir selber auch. Auf keinen Fall sind sie ein Beitrag der Natur, unser Dienstleistungsgewerbe um Vierbeiner zu ergänzen. Sie sind keine Butler, keine Sklaven und auch keine Accessoires zur persönlichen Image-Bildung. Sie sind Hunde. Hund für Hund. Keiner wie der andere. Auch deshalb geben wir ihnen einen Namen, mit dem wir sie anreden. Und Lillys Züchtung war ganz unverwechselbar: Pirosch, der Jagdhund, der die Hasen leben ließ.

Pirosch, the Hound that Let the Rabbits Go

One of the first dogs in my life was Pirosch, a Hungarian Pointer. This was long before I was even aware of the plight facing so many of our dogs. That's why he remains the only dog I've ever had that I picked up at a dog-breeding farm. The name of his breeder was Lilly Wossala. In order to verify his color, blaze and descent, I planned to accept Pirosch from her in person. When I met her, Lilly led me into a room lined with camping cots and she told me that Viszla, the breed's Hungarian name, are simply uncomfortable in baskets, kennels or blankets on the floor.

This was more than 30 years ago and I have to admit, to my embarrassment, that the conformist within me sometimes took over in those days. At that moment, an unforeseeable circumstance made me open my eyes: Pirosch simply wasn't into hunting. On the contrary,

he actually made best buddies with Hansi, my rabbit. He had the kind of character that made it natural for him to appreciate his "flaw." Having had the privilege of sharing 16 years with him, I quickly learned from Pirosch that dogs are individuals just like we are. They are *not* nature's way of supplying us with four-legged field workers. They're not butlers or slaves, nor are they accessories to some personal image enhancement. They are *dogs*. Every single one of them. None are alike. That's one of the reasons we give them names to call them by. And Lilly's breed was as unique as they came: Meet Pirosch, the hunting dog that let the rabbits go.

„Auch wenn es gelänge, die Tiere vor uns zu schützen, wir hätten nichts erreicht. Erst wenn es gelingt, die Tiere nicht mehr schützen zu müssen, sind wir am Ziel. Dann haben wir etwas verändert: UNS."

"Even if we were to succeed in protecting animals from ourselves, we won't really make a difference. Our goal has to be the day when animals will no longer *need* any protection. That's when we'll see real change—in OURSELVES."

Als Benny mich aussuchte

Ob wir einen großen schwarzen, etwa zehn Jahre alten Hund aufnehmen könnten, fragte die Mitarbeiterin eines Altenheims, die bei uns anrief, um zu verhindern, dass der Hund womöglich eingeschläfert würde. In vielen Seniorenstiften werden noch immer keine Tiere geduldet, und so stand nun ein älterer Herr mit seinem Hund ratlos da und wusste nicht weiter.

Gut Aiderbichl ist kein Tierheim, und darüber hinaus wussten wir, dass große schwarze Hunde schwer zu vermitteln sind, noch dazu ältere. Trotzdem nahmen wir uns vor, eine Lösung zu finden.

Benny kam mit seinem Herrchen in einem kleinen Elektroauto zu uns auf das Gut. Stolz auf sich und seinen Besitzer schaute er aus dem Gefährt. Er war der ständige Begleiter seines Herrchens, das sah man deutlich, er wirkte selbstbewusst, wachsam und treu.

Dem Rentner fiel die Trennung genauso schwer wie dem Hund. Benny durfte sein Herrchen doch ein Leben lang begleiten – warum jetzt auf einmal nicht mehr? Es war schon traurig mit anzusehen, und zugleich hatten wir keine Ahnung, wo wir Benny unterbringen sollten. Vielleicht

WHEN BENNY PICKED ME

The question of whether we could take a large, black dog about ten years of age came from a staff member of a retirement home, who'd called us to prevent the dog from possibly being put to sleep. Many retirement centers still have a no-pets policy, leaving an elderly gentleman and his dog with no idea of what to do and where to turn to.

Actually, Gut Aiderbichl is not an animal sanctuary and, moreover, we knew it was hard to find homes for large black dogs, especially older ones. Nevertheless, we were determined to find a solution.

Benny and his master met us on the property in a small electric car. Proud of himself and his owner, Benny peeked out of the vehicle. It was evident that he was his master's constant companion; the dog appeared self-confident, vigilant and loyal.

To separate was just as hard on the retiree as it was on the dog. Benny had spent his whole life at his master's side—how could it suddenly be over? It was quite sad to watch

them and, at the same time, we had no idea where to put Benny up. Suppose he could accompany the night watchman on his patrol of the grounds? However, before we even had a chance to start making arrangements, Benny took advantage of one careless moment and ran off. Hours went by before he was found sitting in front of his master's house door—more than 9 miles from Gut Aiderbichl. This happened twice. Benny wasn't a social dog; he was

würden wir ihn dem Nachtwächter auf seinen Streifzügen über das Gut zur Seite stellen? Aber bevor es dazu überhaupt kommen konnte, war Benny in einem unbeobachteten Moment entwischt und wurde erst Stunden später vor der Haustür seines Herrchens sitzend aufgefunden – über 15 km von Gut Aiderbichl entfernt. Das passierte zweimal. Benny war kein Gesellschaftshund, er war für einen Menschen bestimmt – sein Herrchen. Ihn niemals zu verraten, war er zu allem bereit und verschloss sich allen unseren Angeboten.

Wir versuchten, ein neues Zuhause für Benny zu finden, indem wir sein Bild ins Internet stellten, in den Kalender von Gut Aiderbichl, inserierten – aber es kam nicht einmal eine Anfrage. Und es sollte noch schlimmer kommen. Als Benny spürte, dass er abgegeben werden sollte, versuchte er dies durch Kostproben seiner Persönlichkeit und seiner Fähigkeiten zu verhindern. Zu einigen Pflegern hatte er schon ein gutes Verhältnis, aber die konnten ihn nicht mit nach Hause nehmen. Also wartete er, bis sie Dienst hatten und zeigte, was er kann. Er wollte sich nützlich machen und begann, das Gut vor Eindringlingen zu beschützen – am unangenehmsten war ihm der Briefträger. Als er ihn zweimal gebissen hatte, musste Benny der Polizei vorgeführt werden.

Was sollte nun geschehen? Benny konnte nicht auf dem Gut bleiben und im Privathaus lebte schon der kleine dominante Burschi. Deshalb kam ich mit meinem Tierpfleger, Hans Eder, überein, dass Benny ins Nebenhaus einziehen und von ihm betreut werden sollte.

Hans Eder kümmerte sich rührend um ihn. Benny hatte den besten Tierpfleger der Welt. Aber wenn er an Hans Eders Hose schnupperte, wusste er, dass er nicht dessen einziger Liebling war. Doch mit der Zeit begann er zu verstehen, baute langsam seine Aggressionen ab und wurde ein glücklicher Hund. Schließlich wurde Benny nach Burschis Tod dessen Nachfolger im Haupthaus. Er zog in mein Schlafzimmer ein, zu den Hündinnen, die er akzeptierte.

Ich selbst war nicht unbedingt sein Fall. Zu abgelenkt, oft nicht da, zu exotisch und rastlos verglichen mit den geruhsamen Erfahrungen aus dem ersten Teil seines Lebens. Aber trotzdem schenkte er mir seine Nibelungentreue und ließ sich nach und nach sogar auf Zärtlichkeiten ein. Das Schönste war für ihn Autofahren. Er wollte immer mitkommen, wie damals im kleinen Elektrowagen.

there for one person—his master. Come hell or high water, Benny would never ever abandon him, and any attempt on our part to persuade him otherwise was futile.

We tried to place Benny by publishing his picture on the Internet and inserting an ad for him in the Gut Aiderbichl calendar without getting a single response. And we hadn't even seen the worst yet. Sensing that he was being put up for adoption, Benny tried to derail the process by showing off exhibits of his character and abilities. Although he already was on good terms with some of the animal caretakers, none of them were able to put him up at one of their places. So he would wait until their shifts started to show them what he was made of. Benny also wanted to prove himself useful and began protecting the property against intruders—the mail carrier being at the top of his list. After he'd bitten the man on two occasions, the police had to step in and take Benny away.

Now what? Keeping Benny at the home was out of the question and my own house was already home to little, dominant Burschi. So I reached an agreement with one of my best animal caretakers, Hans Eder, to put Benny in the house next door and under his care.

Hans Eder took tender care of him. Benny had the best animal caretaker in the world. But when he sniffed at Hans Eder's pants, he realized he wasn't the caretaker's only pet. As time passed, however, he learned how to deal with it, gradually overcoming his aggressions and becoming content. After Burschi's death, Benny eventually became his successor in the main house. He found a new home in my bedroom among the female dogs, accepting them all.

He couldn't quite warm up to me though. I guess I was just too distracted, often

Nach fünfjähriger Trennung besuchte uns sein ehemaliger Besitzer auf Gut Aiderbichl. Ich rechnete damit, dass mich Benny verlassen würde. Als der Moment der Wahrheit kam, sprang er seinem früheren Herrchen begeistert entgegen und begrüßte ihn leidenschaftlich. Ich ließ die beiden allein und setzte mich auf eine Bank um nachzudenken. Und wie ich so dasitze, springt mich freudig ein Hund an und legt sich neben mich. Benny! Er hatte sich für mich entschieden. Gemeinsam gingen wir zu seinem Vorbesitzer und verabschiedeten uns herzlich.

Ein halbes Jahr, bevor Benny mich endgültig verlassen sollte, hatte er Probleme mit der Sauberkeit. Daraufhin entfernte ich die Teppiche, sodass einem glücklichen Zusammen-

leben nichts mehr im Wege stand. Auch wenn ich des Öfteren in der Nacht aufstehen musste, um aufzuwischen und anschließend nicht wieder einschlafen konnte. Dann legte ich mich auf mein Bett, sah die schönen Bilder an den Wänden und die Vorhänge aus einer Zeit, als ich noch auf Gestaltung und Farbkombinationen großen Wert legte. Und wie sah das Haus mit all den Tieren jetzt aus?

In diesen Nächten sagte ich mir: Ich will. Ich will genau das. So will ich leben. Bis zu meinem letzten Atemzug. Das macht mich glücklich.

absent, too exotic and too restless compared to his complacent existence during the first part of his life. And yet, he extended his phenomenal loyalty to me as well, gradually even opening up to small tokens of affection. His ultimate joy was to ride in cars. He always wanted to come along, just like in that small electric car back in the day.

After five years of separation, his former owner came to visit us at Gut Aiderbichl. I anticipated that Benny would leave me. When the moment of truth came, he enthusiastically ran up to his old master to give him a glowing welcome. I left the two of them alone and sat down on a bench to reflect on the situation. And as I'm sitting there, this dog joyously jumps up on me and then lies down beside me. Benny! He had chosen me. Together, we walked up to his former owner and warmly bade our farewells.

Half a year before Benny was to depart from my life forever, he started having hygiene problems. So I got rid of my carpets in order to render our happy coexistence complete. It didn't matter if I had to get up several times a night to do a little cleanup and not be able to sleep afterward. On moments like that, I'd lie down on my bed and I'd see these pretty pictures on the walls and these curtains that were all remnants from a time when I greatly cared about design and color combinations. Look at the house now, with all these animals in it!

On those nights, I'd tell myself it's my choice. This is exactly what I want. This is how I want to live. Right up to my last breath. This is what makes me happy.

Baby, Schutzengel der Unterdrückten

Die Berner Sennenhündin Baby wurde auf Gut Aiderbichl am Zaun einer Box angebunden. Dazu ein Abschiedsbrief, der nicht sehr aufschlussreich war. Nicht einmal ihr Name wurde darin erwähnt. Aber sie schien niemanden zu vermissen, und es war, als hätte Baby selbst alles geplant. Sie sah mich an und beschloss: „Du bist ab jetzt mein Herrchen."

Ich behielt sie, obwohl sie das Naturell eines Dragoners hat. Kleider zum Wechseln sollten stets bereit liegen: Zur Begrüßung springt sie einen an, vehement und temperamentvoll, besabbert die Hose. Mit ihrem riesigen Schwanz wedelt sie dann noch ein paar Gläser und Vasen vom Regal. Eigentlich zeigt sie damit, was alles überflüssig ist, und dass auch leere Wohnungen schön sein können.

In zwei Fernsehsendungen über Gut Aiderbichl spielt Baby eine Hauptrolle, sie begleitet die Moderatoren durch die Sendung – und wurde zum Alptraum der Damen vom Kostüm.

Zuhause ist Baby ganz anders. Sie hat das sanfte Gemüt, das sich oft hinter rauer Schale verbirgt. Und deshalb schließen sich ihr auch die kleinen Hunde an. Sie ist die Mama der Unterdrückten und wirft sich für sie in Pose, wenn sie glaubt, dass Gefahr droht.

Mir fallen nicht viele Menschen ein, die einen so großen Hund mit einem so wilden Temperament halten können. Deswegen bin ich froh, dass sie den Weg zu mir gefunden hat, mein „Riesenbaby".

Hallo
ich bin ein Berner
Sennenhund - Weibchen,
3 Jahre alt, und suche
ein neues Zuhause.
Bin sehr kuschelig und
kinderlieb, brauche
aber ein bisserl Platz.
Mein Herrchen muss ins
Krankenhaus und verschenkt
mich.

BABY, GUARDIAN ANGEL OF THE OPPRESSED

Baby, a female Bernese Mountain Dog, was found leashed to a box fence at Gut Aiderbichl. There was also a farewell letter offering little information. It didn't even reveal the dog's name. Baby, however, didn't seem to be missing anyone. It was as if she'd planned the whole thing herself. She looked at me as if deciding, "You're my master now."

I kept her, even though she had the nature of a dragoon. Having Baby around means you require plenty of extra clothing, her way of greeting visitors is to jump at them in ecstasy and idolatry and to drool all over their pants. Wagging her big tail, she's also apt to swipe a couple of glasses and vases off their shelves. I guess it's her way of pointing out anything we don't really need and that empty homes can be nice too.

Two TV documentaries have been made about Gut Aiderbichl with Baby in the main cast as she accompanied the hosts through the episodes—and became the costume designer's worst nightmare.

At home, however, Baby is nothing like that. She has the kind of soft nature often found inside a tough shell. That's what draws small dogs to her too. She's like the "Guardian of the Oppressed," going into defense mode for them whenever she feels a threat.

I don't know too many people who can handle such a combination of big dog and wild temperament. That's why I'm glad my "big Baby" found her way to me.

Die kleine Lily von der Mülldeponie

Früher war es ganz leicht, einen Hund aus dem Süden Europas oder aus der Türkei nach Österreich, Deutschland oder in die Schweiz mitzubringen, was heute wegen der Gefahr von Seuchen und Krankheiten nicht ohne weiteres möglich ist. Aber damals rettete ich jede Menge Hunde aus der Türkei. Ihnen ging es sehr schlecht dort, was die Folge einer inkonsequenten Behandlung war. Zunächst ließ man sie frei gewähren und wenn sie sich dann vermehrten wurden sie vergiftet, erschossen oder noch schlimmer, in Containern auf Müllhalden gebracht.

Mein Beruf brachte mich oft in die Türkei, und ich stellte Kontakte zu tierlieben Menschen her. Darunter türkische Studenten, Tierärzte und Frauen, die sich weltweit für arme Tiere einsetzen.

Am Flughafen war ich schon bekannt: keine Abreise ohne jede Menge Hundeboxen. In Zürich, Wien oder München nahmen mich Tierfreunde in Empfang, wir brachten die Hunde in Pensionen unter und vermittelten sie an ausgesuchte Halter. Bei meinem letzten Hundetransfer aus Istanbul war eine besonders ängstliche Hündin dabei. Ich wollte sie gleich behalten, um ihr die Wartezeit bis zu einer Vermittlung zu ersparen.

Lily glaubte von diesem Moment an nur noch an mich. Beobachtete jeden meiner Handgriffe und wich mir nicht mehr von der Seite. Sie hatte vor jedem Mann Angst, außer vor mir.

Ich hatte zu dieser Zeit mit dem Bau von Gut Aiderbichl begonnen, und eine Salzburger Fernsehredakteurin, die davon erfahren hatte, suchte mich auf. Wir kamen auch privat ins Gespräch, und sie erzählte mir, dass gerade ihr Hund gestorben sei. Da entdeckte sie Lily, die natürlich nicht auf dem Boden, sondern auf einem Sessel saß.

Ich stellte zu meiner Freude fest, dass Lily keine Berührungsängste hatte und sogar etwas tat, was sie vorher nie gemacht hätte: Sie lächelte. Es war ganz klar, dass sie mich in Kürze verlassen und für immer bei der Redakteurin bleiben würde, die später eine meiner besten Freundinnen wurde.

Wir wollten ganz behutsam vorgehen und beschlossen, dass Lily mich auch ab und zu besuchen kommen darf – was heute noch Tradition ist. Außerdem ist Lily nur ein paar

LITTLE LILY FROM THE LANDFILL

In the old days, it was simple enough to take a dog from southern Europe or Turkey to Austria, Germany or Switzerland. It's a lot harder nowadays, because of the threat of epidemics and diseases. Back in the day, though, I used to rescue scores of dogs out of Turkey. Incoherent management policies made their lives an ordeal there. At first, they were allowed to roam freely, but once they began to propagate, they were poisoned, shot, or worse, put in containers and transported to landfills.

Back then, my job required me to travel to Turkey a lot, and as I did, I also used to meet with animal protectionists. They included Turkish students, veterinarians and women championing the rights of abandoned animals worldwide.

Among airport officials, I was already known for never departing without scores of boxes containing dogs. Arriving in Zurich, Vienna or Munich, I was met by fellow animal lovers and we proceeded to place the dogs in sanctuaries or place them with select keepers. My last dog transfer from Istanbul included one female that was particularly fearful. I wanted to keep her right away in order to spare her the long wait for an adoption.

From that moment on, Lily trusted no one but me. She watched every move I made and never strayed from my side. She was afraid of every man except me.

Around that time, I had started on the construction of Gut Aiderbichl when a journalist for a TV station from Salzburg heard of my project and came to visit me. We also got to talk on a personal level and she told me that her dog had just died. Then her eyes fell on Lily sitting not on the floor but in—you guessed it—an armchair.

I was thrilled to see Lily express no fear of contact and to actually witness her do something she never would have done before—she smiled! It was plain to see that she'd be leaving me soon for her new life with this journalist, who went on to become one of my best friends.

In order to make the transition as easy as possible for Lily, we decided to arrange for her to still visit me sometimes, a tradition we still keep up. In addition, Lily lives just a few blocks from me. But here's the ticker: No matter from which direction we come in, she always knows where she lives. Sometimes she thinks I'm the one who's confused and she'll start barking to set me straight. Without having to peek out of my Jeep, she knows where she is by instinct. What's more, she can read me. For example, when I say to her, "You're staying with me today," she instantly goes quiet and she's pleased as punch. Never once has her intuition failed her.

So far, I was the only one who knew that she came from a landfill. Not that there's any way to tell, because Lily is a decidedly noble dog. Whenever she can, she sleeps on a pillow on a bed. Encountering loud dogs, Lily will turn the other way with her head held up high. And she doesn't refrain from using that same approach on people too.

At the moment I'm writing this, she's lying at my feet, and when I say her name out loud, she wags her tail on the floor. She was brought here by intervention. If not for that, nobody would ever have had the chance to see how special she is.

Straßenzüge weiter gezogen. Das Unglaubliche ist: Ganz egal, aus welcher Himmels-
richtung wir mit dem Auto kommen, weiß sie, wo sie wohnt. Und weil sie glaubt, dass ich
zerstreut bin, bellt sie und erinnert mich. Sie kann aus meinem Jeep nicht hinausschau-
en, sie spürt einfach, wo sie ist. Außerdem versteht sie mich. Wenn ich sage: „Du bleibst
heute bei mir", ist sie sofort still und freut sich. Noch nicht ein einziges Mal hat sie ihr
Gefühl getäuscht.

Dass sie von einer Müllhalde kommt, wusste bisher nur ich. Niemand würde das ahnen,
denn sie ist ein ausgesprochen vornehmer Hund. Wenn irgend möglich, schläft sie auf
einem Kopfkissen auf dem Bett. Lauten Hunden geht Lily, ihr Köpfchen in den Nacken
werfend, aus dem Weg. So verhält sie sich auch Menschen gegenüber.

Sie liegt mir gerade beim Schreiben zu Füßen, und immer, wenn ich ihren Namen sage,
klopft sie mit dem Schwanz auf den Boden. Dass sie hierher kam, war eine Fügung. Bei-
nahe hätte niemand je erfahren, was für ein wunderbares Geschöpf sie ist.

DIE BLINDE TILLI

Ehrlich gesagt zögerte ich, bevor ich mich dazu entschloss, eine blinde Hündin bei mir zu Hause aufzunehmen. Bis zu jenem Tag, als die blinde Tilli aus Bulgarien mit einer Gruppe von ganz armen Hunden zu uns kam. Sie war so gut wie nicht vermittelbar, und deshalb gab ich ihr schließlich eine Chance bei mir.

Wir integrierten sie in einer Gruppe von Hunden, die Hans Eder in meinem Privathaus betreut. Zeigten ihr zuerst die Wohnung, dann den Garten, und wie man ins Auto springt. Dann stellten wir sie den anderen Hunden vor. Als sie nach kurzer Zeit bemerkten, dass Tilli nicht sehen kann, gingen sie von diesem Moment an und bis zum heutigen Tag besonders behutsam mit ihr um. Dabei hört man ständig: „Die Natur ist grausam", und: „Im Hunderudel herrscht eine gnadenlose Hierarchie". Falsch, Tilli wurde zugleich als Spielkameradin und Familienmitglied ebenbürtig von allen akzeptiert.

Die Natur hat es schlussendlich mit Tilli doch noch gut gemeint und ihre anderen Sinne schärfer entwickelt. Die blinde Tilli hört besser, riecht besser, aber vor allem fühlt sie besser als andere Hunde. Das hat sie mehrfach bewiesen.

Einmal im Monat muss ihr Hans Eder eine Lösung ins Ohr träufeln, das hat sie besonders ungern. Wenn die Stunde kommt und er sein Wohnzimmer, in dem sie wohnt, mit der Lotion betritt, springt sie sofort von der Couch und versteckt sich. Das erstaunt selbst den sonst nicht so leicht zu beeindruckenden Hundepfleger, denn an allen anderen Tagen bleibt sie liegen, wenn er kommt.

Dass sie kein Problem im Umgang mit den anderen Hunden hat, ist sicherlich auch ein Verdienst ihres bezaubernden Wesens. Bei Hunden stellt sich außerdem für mich immer mehr heraus: sie verhalten sich, wie man sie hält. Wir interpretieren das Wesen der Hunde ohne Grundlage, wenn wir das nicht bedenken.

Die „Eder-Hundegruppe" folgt täglich dem gleichen Ritual. Frühstück, immer zur gleichen Zeit, und dann fahren sie miteinander auf die Hundefreilaufwiese von Gut Aiderbichl und toben sich aus. Danach gibt es viel Ruhe und noch drei weitere „Gassis". Am Abend noch ein „Betthupferl". Das Geheimnis ihres Wohlbefindens scheint die Routine eines gesicherten Alltags zu sein, und die Gewissheit, dass ihre Erwartungen nicht enttäuscht werden.

BLIND TILLI

To be honest, my decision to take a blind female dog into my home wasn't without its hesitations. That changed on the day that Tilli came to us with a group of very forlorn dogs from Bulgaria. Since it was next to impossible to find anyone willing to adopt her, I finally gave her a chance in my home.

We put her in with a group of dogs cared for by Hans Eder in my private home. First, we showed her the house, then the garden and then we showed her how to hop into a car. Then we introduced her to the other dogs. They soon realized that Tilli couldn't see and,

from that moment on to this very day, they have always been extremely gentle on her. I know there's all this talk about "the cruelty of nature" and the "merciless hierarchies among dog packs." It's not true! Every one of those dogs welcomed Tilli as an equal, not only as a playmate but also as a member of the family.

In the end, nature meant well for Tilli after all by in-tensifying Tilli's remaining senses beyond the usual. Tilli can hear better, smell better and especially feel better than other dogs. She has demonstrated that on many an occasion.

Einer meiner noch nicht erfüllten Träume ist, ein Hundedorf nach unserem Haltungs-schema zu bauen. Anstelle von Zwingern oder Betonbuchten gäbe es Pfleger, die in Häu-sern mit Hundegruppen mit bis zu 20 Hunden leben können. Ein lebenswertes Leben auch für die Tiere, die schwer vermittelbar oder behindert sind. Was man dann noch bräuchte, wären Pfleger wie Hans Eder, durch und durch integer und immer das Wohl der Hunde im Sinn. Täuschen lassen sich die Tiere sowieso nicht. Man denke nur an Tilli und die übersinnliche Wahrnehmung der Ohrentropfen.

Once a month, Hans Eder has to drip some lotion into Tilli's ears—a procedure she hates with a passion. Whenever it's that time and he enters the living room where she lives with the lotion in his hand, she instantly jumps off the couch and hides somewhere. Although he has seen a lot in his days as a dog trainer, even Hans Eder is baffled by it, because on any other day, she just remains in place whenever he comes in.

Her ability to get along so well with the other dogs is without a doubt part of her charming nature. But there's something else as well that I see more and more in dogs: they behave the way they're treated. Failing to take that into account, we lack any foundation for interpreting a dog's character.

Every day the "Eder Dog Group" follows the same ritual. After breakfast, which is always at the same time, the whole gang heads out to the off-leash area of Gut Aiderbichl to have a good romp. This is followed by plenty of naps and three more "walks." Then there's a "nightcap" in the evening. The secret to the dogs' well-being appears to be the routine of a stable everyday life as well as the knowledge that they won't be let down in their expectations.

One of my unfulfilled dreams is to build a dog village based on our training principles. Instead of kennels or concrete barriers, I envision caretakers living in houses with groups containing as many as 20 dogs. Talk about a life worth living even for animals that are hard to be placed or handicapped! Of course, I'd also need trainers like Hans Eder, people who are all-out reliable and focus on the well-being of the dogs. Besides, these animals won't be fooled, anyway. Just think of Tilli and her uncanny way of anticipating those ear drops.

MEINE SCHLAFZIMMER
HUNDE HEUTE

MY BEDROOM DOGS
TODAY

LUCYS KUMPELHAFTE NACHSICHT

Sie kam aus einem Tierheim auf Mallorca. Freunde nahmen sie von dort mit nach Salzburg. Aber die Haltung von Hunden mit trauriger Vorgeschichte ist schwierig, und so kam Lucy letztlich zu mir.

Sie gehört zur Rasse der Spanischen Jagdhunde, für die es in Spanien eine recht merkwürdige Vorbereitung auf ihre Aufgaben gibt. Man lässt sie einfach ein paar Wochen hungern, damit sie ganz schnell die Fährten zu Hasen und Fasanen finden. Das erklärt Lucys unstillbaren Hunger bis heute. Wenn man nicht aufpasst, ist ein Teller blitzartig abgeräumt und wieder ein Stück Kuchen verschwunden. Sie hat sich ein Fettpolster zugelegt, sicherheitshalber, für den Fall, dass man sie wieder hungern lässt.

Sie muss in einem Dorf mit lauten Kirchenglocken gelebt haben. Wenn sie mit mir einen Film sieht, in dem Glocken läuten, dreht sie fast durch. Sie erinnern sie offenbar an früher.

Obwohl Lucy von allen meinen Hunden derzeit am längsten bei mir lebt und geschätzte 8 Jahre alt ist, verhält sie sich Neuankömmlingen gegenüber tolerant. Sie tritt selbst ihr Körbchen ab. Aber wenn es zum Füttern kommt, dann gibt es kein Pardon. Dann geht es um die Wurst.

Wäre Lucy ein Mensch, sie würde anderen ihre Hilfe nicht versagen. Sie hat fast etwas Kumpelhaftes und lebt gerne „miteinander". Vielleicht sagt deshalb Laura, die Tochter meines Geschäftsführers Dieter Ehrengruber: „Die Lucy ist meine Schwester."

Lucy's Sociable Patience

She came from an animal shelter in Majorca. From there, friends of mine took her with them to Salzburg, Austria. However, it's hard to train a dog with a sad past, so Lucy eventually came to me.

She belongs to the breed of Spanish hunting dogs and, in their original country, these dogs are prepared for their jobs in a quite barbarian way. They are simply starved for a couple of weeks so they're driven to hunt hares and pheasants with extreme voracity. That would explain the insatiable hunger Lucy keeps exhibiting to this day. Turn your back to her and, in no time at all, a plate has been emptied and another piece of pie has vanished. She has put on a fat pad just in case anybody decides to starve her again.

She also must have lived in a village with loud church bells. Anytime we watch a movie containing the ringing of bells, she almost goes ballistic. It seems to remind her of her past.

Although Lucy has lived with me longer than any of my other dogs have, and she's an estimated 8 years of age, she's tolerant of new arrivals, even offering them her basket.

But when it's feeding time, she knows no mercy. It's all or nothing to her.

If Lucy were human, chances are she'd always be there for others. There's an almost sociable quality to her and she likes the concept of "together." Maybe that's why Laura, the daughter of my business manager, Dieter Ehrengruber, always says, "Lucy is my sister."

Ricki und die Cousine dritten Grades Lulu

Gemeinsam mit 69 Artgenossen und 19 Katzen saßen Ricki und Lulu in einem von der Polizei beschlagnahmten Transport von Ungarn nach Berlin. Ricki war, wie all die anderen, außergewöhnlich niedlich, und es hätte keinen Grund gegeben, dass ich sie zu mir nach Hause hole. Aber Lulu, ihre Cousine dritten Grades, ist bewegungsbehindert und manchmal desorientiert. Es war mir ein zu großes Risiko, sie zu vermitteln, also behielt ich gleich beide.

Ricki und ihre Cousine sind blitzgescheit und machen ständig etwas ausfindig, das sie zerkauen können: Telefonkabel, Kugelschreiber, Feuerzeuge und sogar echte Euroscheine. Von Lulu und Ricki habe ich Ordnung gelernt. Seit sie hier sind, sind die Flächen meiner Privaträume so aufgeräumt, als würde ich im Hotel leben. Unfasslich, dass es Geschöpfe gibt, die nur wenige Kilogramm wiegen und so intelligent und sensibel sind wie Ricki und Lulu.

Ricki and her Third Cousin, Lulu

Along with 69 other dogs and 19 cats, Ricki and Lulu were part of a transport headed from Hungary to Berlin when authorities confiscated it. Like all the others, Ricki was highly adorable and it wouldn't really have been necessary for me to take her in. Except that Lulu, her third cousin, is physically handicapped and occasionally becomes disoriented. I felt I was taking chances by placing her, so I just kept both of them.

Ricki and her cousin are both exceptionally bright and always find something they can chew to bits: phone cables, pens, lighters and even Euros. Lulu and Ricki have taught me order. Ever since they've been here, I keep every surface in my home as neat as if I was living in a hotel. It's hard to imagine creatures weighing just a few pounds and possessing the same level of intelligence and sensitivity that's evident in Ricki and Lulu.

Kantor, den keiner wirklich kannte

Eine österreichische Urlauberin hatte in einem ungarischen Dorf beobachtet, dass ein Schäferhund seit einigen Tagen allein ein Haus bewohnte, weil sein Besitzer gestorben war. Irgendjemand versorgte ihn, mehr oder weniger, ab und zu.

Sie sei sehr beunruhigt, sagte sie mir, und dass ihr sein ungewisses Schicksal nicht mehr aus dem Kopf ginge. Ich traf eine Vereinbarung mit der tierlieben Dame: Wenn sie nach Ungarn fährt und den Hund bekommt, übernehme ich die Verantwortung für sein künftiges Leben. Eine Entscheidung, die mir nicht leicht fiel. Schäferhunde sind auch in Österreich und Deutschland schwer vermittelbar und in meinem Privathaus habe ich schon zu viele Hunde.

Schon am nächsten Tag kam die Dame mit Kantor nach Gut Aiderbichl in Henndorf. Dort wurde er erstversorgt und beobachtet. Unsere Tierpfleger machten sich Sorgen. Der damals etwa 6-7 Jahre alte Schäferhund zeigte Aggressionen und stürzte wie wild hinter einer Katze her. Also empfahlen sie mir eine baldige Vermittlung an einen guten Platz. Wie durch ein Wunder meldete sich zu unserer großen Freude ein pensionierter Hundeführer, der

bereit war, zunächst einmal auf Probe ein gemeinsames Leben mit Kantor zu führen.

Doch einige Wochen später rief er mich verzweifelt an. Kantor sei so anders als die typischen Schäferhunde, die er in seinem Leben hatte, vor allen Dingen sei er aggressiv und er leide an der Hüftkrankheit, die viele Schäferhunde im Alter trifft, ein Problem der Züchtung. Sein Haustierarzt empfahl aus Gründen des Tierschutzes die Einschläferung.

Ich fand, dass Kantor eine letzte Chance bekommen muss, und holte ihn zu uns zurück. Mein privater Tierpfleger Hans Eder und ich beschlossen, ihn vorerst in meinem Büro

KANTOR, THE MYSTERIOUS ONE

While vacationing, an Austrian lady came into a Hungarian village where she noticed how a German Shepherd had been spending several days in a house all by himself, because his owner had passed away. The dog seemed to have someone looking after him—sporadically, anyway.

The lady told me that she was very troubled by what she saw and that she was very concerned about the dog. So I made an arrangement with her: If she were to return to Hungary to pick the dog up, I would assume responsibility for his future life. It was a decision that didn't come easy. German Shepherds are tough to place in Germany and Austria, too and my private home already housed too many dogs.

The lady arrived with Kantor at Gut Aiderbichl in the village of Henndorf the very next day. There we gave him a quick examination and put him under observation. Our animal caretakers were worried about him. The shepherd dog, maybe 6-7 years old at that point, showed signs of aggression and charged after a cat as if he'd gone berserk. I was told he needed to be placed in a good home A.S.A.P. Miraculously, and to our great relief, a retired dog trainer contacted us saying he'd be willing to take Kantor in on a trial basis.

A few weeks later, however, I received a desperate phone call from him. He told me that Kantor was radically different from the typical shepherd dogs he'd had in his life. Above all, he pointed out the dog's aggressive behavior and an ailment involving his haunch, a symptom found in many older shepherd dogs due to a breeding problem. Citing animal safety concerns, his veterinarian recommended the dog be put to sleep.

I felt that Kantor deserved a final chance and took him back. My private animal caretaker, Hans Eder, and I decided to temporarily keep him in my office, since the main building already accommodated too many animals, and Kantor obviously had a problem with

that. We only let him out into the garden and installed safety gates in our office building to help him climb the stairs without pain.

Then someone forgot to shut the door one day. I just froze when I saw Kantor stand amidst four of my female dogs and braced myself for the worst.

Instead, Kantor let out a cheerful yap, extended his forelegs and they all started chasing and playing around and licking each other in a game of good spirits, the likes of which I'd rarely witnessed before. There was no sign of a haunch problem either.

A few days later, we introduced Kantor to our males, too. Not a problem! He just applied his Hungarian charm. Only a small, cranky Chihuahua adopted by a colleague after it lost its placement remains at odds with Kantor. Yapping loudly, Scubi will jump at Kantor and almost reach his shoulders. Kantor, however, takes it in stride. In view of Scubi's miniature size, Kantor simply wouldn't think it's fair to take him on.

On evenings, all my dogs are allowed a nightcap and they all form a single-file line to face me. Kantor waits his turn without pushing. Maybe he views this moment as a special occasion and he simply wants to look his best.

I have also discovered that Kantor is an excellent soccer player. It's too bad that many opportunities to talk to previous owners of the dogs we receive are lost. Thus, it often takes a lot of time to find out what makes each dog special and to establish a relationship with him or her.

Perhaps Kantor's deceased owner is watching us from above, now as grateful for this happy end as we all are. To me, the story of this shepherd dog proves once again that a death sentence doesn't have to spell out death and that we must never give up hope.

Kantor is now roughly 10 years old and lying on the floor next to my feet as I'm writing this. His presence is a blessing. He has made me wiser and more considerate. I'll be grateful to him for the rest of my life.

unterzubringen, denn im Haupthaus leben schon viele Tiere, und damit hatte Kantor ja offensichtlich ein Problem. Wir ließen ihn nur im Garten laufen, und damit er keine Schmerzen beim Treppensteigen hatte, bauten wir im Bürohaus Kindersperren ein.

Dann vergaß eines Tages jemand, die Tür abzuschließen. Ich erstarrte fast zur Salzsäule, als Kantor plötzlich zwischen vier meiner Hündinnen stand, und rechnete mit dem Allerschlimmsten.

Doch Kantor stieß einen Juchzer aus, spreizte seine Vorderbeine, und ein Spiel der Freude mit Jagen und Wälzen und gegenseitigem Ablecken der Lefzen begann, wie ich es selten vorher gesehen habe. Auch ein Hüftleiden war nicht zu bemerken.

Wenige Tage später stellten wir Kantor auch unsere Rüden vor. Kein Problem! Er setzte seinen ungarischen Charme ein. Nur ein kleiner grantiger Chihuahua, den eine Mitarbeiterin aufnahm, weil er seinen Platz verloren hatte, ist gegen Kantor. Scubi springt laut kläffend bis auf Schulterhöhe an Kantor hoch. Doch Kantor lässt ihn gewähren, schließlich ist Scubi sehr klein, und so würde er es unfair finden, ihm etwas anzutun.

Am Abend bekommen alle meine Hunde ein Betthupferl. Wie Orgelpfeifen sitzen sie dann vor mir. Kantor wartet und drängelt nicht. Er empfindet das wohl als feierlichen Augenblick und will dann ganz besonders lieb sein.

Noch etwas habe ich herausgefunden. Kantor ist ein exzellenter Fußballspieler. Schade, dass wir häufig nicht mehr mit den Vorbesitzern sprechen können, wenn uns Tiere gebracht werden. So dauert es oft eine ganze Weile, bis man Besonderheiten heraus- und anschließend zueinander findet.

Vielleicht sieht uns Kantors verstorbener Vorbesitzer von oben zu und ist jetzt für das Happy End genauso dankbar, wie wir alle. Die Geschichte dieses Schäferhundes hat mir einmal mehr gezeigt, dass ein Todesurteil kein Todesurteil sein muss, und man die Hoffnung nie aufgeben darf.

Kantor ist jetzt geschätzte 10 Jahre alt und auch er liegt neben mir auf dem Boden, während ich schreibe. Ein Glück, dass es ihn gibt. Er hat mich weiser und rücksichtsvoller gemacht. Ich werde es ihm mein ganzes Leben lang danken.

JEANNY, DAS GLÜCK AUF VIER BEINEN

Jeannys Vorbesitzer, ein Landwirt, hatte sich verschätzt als er aus der Jack-Russel-Hündin einen Haus- und Hofhund machen wollte. Sie war den ganzen Tag im Wald unterwegs, und die Familie musste ständig nach ihr suchen. Dann brachte man sie zu uns.

Es vergingen Monate, bis sie sich angesprochen fühlte, wenn jemand sie rief. Inzwischen aber benimmt sie sich vorbildlich und fair gegenüber Menschen und Tieren. Ich glaube daran, dass die beste Erziehung auf Geduld und Logik basiert, was sich auch bei Jeanny bestätigte.

Wenn sie die Vergangenheit einmal einholt, sie ein riesiges Loch im Garten gräbt und anschließend stolz und völlig verschmutzt vor mir steht, dann lobe ich sie. Das entspricht ihrem Wesen, es macht einen großen Teil ihrer Persönlichkeit aus. Sie läuft jetzt nur noch im Schlaf weg. Da liegt sie in ihrem Körbchen, bewegt die kleinen Beine und bellt halblaut. Dann ist sie die Jeanny von früher, die mir allerdings keine Minute Sorge bereitet.

Als ich schon dachte, ich hätte genug von Jeanny berichtet, kam Hans Eder zu mir und erzählte, dass sie das Schwimmen entdeckt hat. Sie ist fasziniert davon, welche Ringe sich im Wasser um sie bilden und auch von den Luftblasen. Ich lief gleich mit der

Jeanny, a Four-Legged Bundle of Joy

Jeanny's previous owner, a farmer, had made a mistake when he decided to train this Jack Russell female as a house- and watchdog. She spent everyday out in the woods, sending the farmer and his family on a constant search for her. So they brought her to us.

Months went by before she even reacted to anyone calling her. Today, however, her behavior is exemplary and she is fair to both people and animals. I believe that the best training is founded on patience and common sense, and Jeanny serves to prove my case.

When her past occasionally catches up with her, she'll dig a huge hole in the garden and stand before me covered with pride and dirt—and I'll praise her. Because that's what's in her genes, forming a major part of her personality. Today she only runs away in her dreams. That's when she'll lie in her basket, move her little legs and emit semi-loud barks. This is Jeanny reverting to her former self, which I've never had a moment's trouble with.

Just when I thought I'd said all there was to say about Jeanny, Hans Eder came up to me, saying she's developed a passion for swimming. She's fascinated by the rings the water forms around her and by the air bubbles it produces. I immediately grabbed my

Kamera zum kleinen Weiher im Garten, und als ich Jeanny lobte, sprang Ricky, ihre beste Freundin, auch gleich ins Nasse. Jetzt ziehen sie zu zweit ihre Kreise. Die Welt hat schöne Seiten.

camera and hurried out to the small pond in the garden. And as I was busy praising Jeanny, along came Ricky, her best friend, and jumped right in too. They now draw their circles in the water together. Life *can* be good.

Snoopys gebrochenes Herz

Es gibt Menschen, die können einen so von unten herauf anschauen, mit einer Eindringlichkeit, dass es einen umschmeißt. Denn diese Blicke sagen: Schaut mir in die Augen! Das ist oft das Schwerste, und die Gründe dafür, dass es schwer ist, sind meist wenig schmeichelhaft. Eine Frau mit so durchdringenden Blicken war Prinzessin Diana.

Aber auch die Blicke von Tieren können uns entwaffnen. Snoopy, eine Beagle-Hündin, hieß Canberra, bevor sie zu mir kam. Sie wurde für ein Versuchslabor gezüchtet und verbrachte dort die ersten zwei Jahre ihres Lebens. Als sie mich zum ersten Mal anschaute, war ich fassungslos. In ihren Augen leuchtete ihr gebrochenes Herz.

Ihr Leben verdankt sie in erster Linie der Organisation „Kölner Modell", die sich mit einem Versuchslabor arrangiert hat und nach Ablauf der Experimente Laborbeagle übernehmen darf. Von diesen suchte ich die ängstlichste Hündin aus, denn ich habe Erfahrung mit Versuchshunden. Mein vorletzter Beagle hieß Snoopy, weshalb ich Canberra umtaufte, ihm zum Gedächtnis.

Diesen Tieren wird die schönste Phase des Lebens geraubt: Das Erwachsenwerden. Neugier – worauf denn? Freude – worüber? Lernen – was denn, außer das grausige Leben aus-

Snoopy's Broken Heart

Some people have a way of looking at others from within with the kind of forcefulness that sweeps people off their feet. Here is a gaze that says, "Look into my eyes!" It can be one of the hardest things to do and the reasons for that are usually less than flattery. One woman who had this powerful gaze was Princess Diana.

However, the gaze of an animal can have that disarming effect on us too. Snoopy, a female beagle, was called Canberra, before she came to me. She'd been bred for a test laboratory, where she had spent the first two years of her life. The first time I caught her gaze, I was stunned. Her eyes revealed a broken heart.

Basically, she owed her life to an organization called "Kölner Modell," or "Cologne Model," which has an arrangement with a certain test lab, in which they hand their test beagles over to this organization once those beagles are no longer used for tests. That was also where I picked up this timid female because of my experience in handling test dogs. My second-to-last beagle had been named Snoopy, which is how I renamed Canberra, in homage to him.

These animals are robbed of the best part of their lives—the part of growing up. Curiosity—why? Joy—what for? Learning—what's there to learn, other than how to cope with an abysmal existence? All that Snoopy ever received from these people over the course of two years was dried foods and water, served with wire mesh and test treatments.

It took at least a year before Snoopy understood the concept of stairs. The sun, the wind, the meadows: they were all new to her. It wasn't until summer 2008 that she jumped into my arms for the first time when I returned from a trip. So I played with her and she licked

zuhalten? Was die Menschen Snoopy über zwei Jahre zugestanden, waren Trockenfutter, Wasser, dazu Gitter und Versuchsbehandlungen.

Mindestens ein Jahr hat es gedauert, bis Snoopy verstand, was Treppen sind. Sonne, Wind, Wiesen – alles war neu. Erst im Sommer 2008 sprang sie zum ersten Mal auf mich zu, als ich von einer Reise zurückkam, wir haben gespielt, sie hat mein Kinn geleckt. Gerührt habe ich ihr meine Lippen auf die Augen gepresst. Das machen Mutterhündinnen bei Welpen, die sie ganz besonders schätzen. Ich konnte ihr allmählich das Geraubte zurückgeben.

Wenn man wenigstens als ersten Schritt offensichtlich überflüssige Tierversuche einstellen würde! Ich habe Snoopy versprochen, dass ich mich, solange ich lebe, dafür einsetzen werde und hoffe, sie spürt, dass sie mir vertrauen kann.

In Snoopy lebt das Andenken an meine Versuchsbeagle, die bereits nicht mehr unter uns sind: Ricki, Snoopy, der Erste, Angy und die vielen, die ich mit Hilfe tierlieber Menschen weitervermitteln konnte.

my chin. Deeply touched, I put my lips to her eyes. That's what female dogs do with pups that they're particularly fond of. Gradually, I was able to give back what had been taken from her.

If only somebody would make the first move by at least banning the use of animals for test purposes that are clearly obsolete! I have promised Snoopy that I will fight for this for as long as I live and I hope her instincts are telling her that she can depend on me.

Snoopy keeps alive the memory of all my lab beagles that are no longer with us: Ricki, Snoopy the First, Angy, and the many others I was able to place in good homes with the help of other animal lovers.

DIE ZWEITE CHANCE IM LEBEN

A SECOND CHANCE
AT LIFE

TARI UND PACO

Es war im Herbst letzten Jahres. Ich arbeitete alleine im Büro, als mich der Anruf einer verzweifelten Dame erreichte. Sie erzählte mir von ihrem Schicksal und dass sie unheilbar krank sei. Damit habe sie sich abgefunden. Nur unter der Ungewissheit, was aus ihren beiden Hundekindern Tari und Paco werden sollte, litt sie unendlich.

Ich fühlte mich, der Situation entsprechend, hilflos und versprach ihr ein sofortiges Fax, in dem ich mich bereit erklärte, die beiden Hunde, sollte ihr etwas zustoßen, bei mir privat aufzunehmen.

Monate vergingen. Dann kam wieder ein Anruf Tari und Paco betreffend, dieses Mal von einer Hundepension. Die beiden waren dort, und mit ihnen mein Schreiben. Ich schickte noch am selben Tag Raimund, einen Mitarbeiter, nach Deutschland zu der Pension.

Am Abend zogen Tari und Paco für immer in mein Haus. Sie waren sehr ängstlich und auch heute noch, nach einem halben Jahr, sind sie vorsichtiger als andere Hunde.

Einige Zeit später erhielt die gemeinnützige Gut-Aiderbichl-Stiftung ein Testament. Ein kleines

TARI AND PACO

It happened in autumn of last year. I was working alone in my office when I received a call from a desperate lady. She told me about her terminal illness and that she had accepted her fate. What really made it unbearable for her was the uncertainty of what was to become of her two puppies Tari and Paco.

I felt as hopeless as anybody would in such a situation. So I promised to send her a fax right away stating my offer to put up her two dogs in my private home, in case something were to happen to her.

Months went by. Then there was another phone call regarding Tari and Paco, this time from a dog sanctuary. They said they had the two dogs, as well as my correspondence. That same day, I dispatched Raimund, a member of my staff, to the Germany-based sanctuary.

That evening, Tari and Paco moved into my home for good. They were very timid and even today, after six months, they still remain more cautious than other dogs.

Some time after that, the non-profit organization of Gut Aiderbichl received a last will. It was a small

Vermächtnis zu Gunsten der Tiere. Die Dame hatte außerdem verfügt, dass die beiden Hunde eingeschläfert werden sollten, es sei denn, sie dürften gemeinsam zu mir. Eine Trennung würden sie nicht überleben.

Noch bevor uns das Vermächtnis erreichte, verlief bereits alles dem letzten Wunsch gemäß. Tari und Paco sind zusammen, sie schlafen ganz nahe bei mir und sind glücklich. Sie verdanken ihr Glück der Weitsicht ihrer Hunde-Mutti und einem Fax, das nicht verloren ging.

bequest benefiting the animals. In her will, the lady had also requested that her two dogs be put to sleep, unless I accepted them together. The idea of separating them was something she simply couldn't bear.

Her last wish had already been fulfilled before we had even received her last will. Tari and Paco are together; they sleep very close to me and they're happy. They owe their happiness to the foresight of their "Mom" and to a fax that didn't get lost.

JACKY

Jackys Frauchen leidet unter Depressionen und weiß sich oft selbst nicht zu helfen. Als sie bemerkte, dass natürlich auch Jacky unter diesen Umständen leidet, entschloss sie sich aus Liebe zu einem schweren Schritt. Sie ließ uns wissen, wenn Jacky nach Aiderbichl kommen dürfte, könnte sie sich mit dem Gedanken anfreunden, sich von ihrem Hund zu trennen.

Jackys Besitzerin stand ein langer Klinikaufenthalt bevor – da durften wir ihn holen. Der Kleine liebt große Hunde, besonders Baby, die Big Mama unserer Kleinhunde. In ihrer Nähe lässt es sich gut mutig sein.

Jacky

Jacky's mistress suffers from depression and often doesn't quite know how to fend for herself. Upon realizing that Jacky obviously suffered under these circumstances too, it was out of love that she made a hard decision. She indicated to us that if Jacky could come to Aiderbichl, she could buy into the idea of parting ways with her dog.

When Jacky's owner was about to face a long stay at a clinic, she authorized us to come pick him up. The little guy is very fond of big dogs, especially of Baby—"Big Mama" to our small dogs. Being around her is a good time to act tough.

MICHI UND MELINA

Auf dem Rückweg von einer Esel- und Hunderettung auf der Insel Thassos in Nord-griechenland, entdeckten wir kurz vor Kavala zwei streunende Hunde. Wie sich später herausstellte, handelte es sich vermutlich um Mutter und Sohn. Wir hatten noch Platz in einem unserer Autos. Vorsichtig lockten wir sie an. Da wir keine Papiere hatten, mussten wir in Thessaloniki eine Tierpension und einen Tierarzt finden. Die beiden konnten uns an diesem Tag nicht begleiten, sondern mussten in Quarantäne. Einige Wochen später holten wir sie nach Gut Aiderbichl. Sie sind jetzt glücklich, dass sie für immer bei uns bleiben dürfen – zusammen für den Rest ihres Lebens.

MICHI AND MELINA

Returning from a rescue mission of donkeys and dogs on the island of Thassos in northern Greece, we came across two stray dogs just outside of the city of Kavala. As it turned out later, the two were likely mother and son. We still had room left in one of our cars. Carefully, we encouraged them to come up closer. Being short of the proper forms, we had to stop in Thessaloniki to find an animal sanctuary and a veterinarian. Instead of coming with us that day, the two dogs wound up being quarantined. Several weeks later, we transferred them to Gut Aiderbichl. Today they're happy they can stay with us forever—together for the rest of their lives.

Gina und Arco

Verzweifelt rief mich vor einem Jahr die Leiterin eines Tierheims bei München an. Die Polizei hatte bei einer Razzia mehrere Kampfhähne und zwei Kampfhunde beschlagnahmt. Ergebnis einer Durchsuchung im Spielermilieu, von dem unsereiner zwar nichts weiß, das aber offenbar gleich um die Ecke seine Unsitten betreibt, z. B. Tierkämpfe, bei denen viele qualvoll sterben. Für die Zocker stellen Kampfhähne und Kampfhunde einen großen finanziellen Wert dar. Da ist viel Geld im Spiel, weshalb sie dann auch versuchten, in das Tierheim einzubrechen und sich die lebenden Wertgegenstände zurückzuholen.

Gemeinsam beschlossen wir, sie an sichere Orte zu bringen und vor ihren Peinigern zu verstecken. Einem der Hunde, Michael, hatte man offensichtlich mit einem Teppichmesser die Ohren abgeschnitten. Damit versucht man, die Tiere aggressiver zu machen und die Kampfdauer zu verlängern. Der Hund mit verstümmelten Ohren steht noch heute unter meinem persönlichen Schutz und lebt an einem Ort, den nur ich selber kenne. Der andere Hund konnte an einen guten Platz weitervermittelt werden.

Auch Gina, die jetzt eine Aiderbichlerin ist, hat man die Ohren abgeschnitten. In Bulgarien, wo sie herkommt. Aus Spaß oder aus ähnlichen Gründen wie bei dem Kampfhund. Vielleicht war es aber auch ein amateurhafter Versuch, ihr die Ohren zu kupieren, man weiß es nicht. Etwas konnten ihr diese Sadisten allerdings nicht nehmen: ihren Charme, ihr wunderbares Wesen, das auch die anderen Hunde erkennen.

Ihr bester Freund ist Arco, der auch aus Bulgarien stammt und vor allen Menschen, außer vor uns und seinem Pfleger, Todesangst hat. Wir haben dafür gesorgt, dass er unter uns bleibt, sicher vor Fremden.

GINA AND ARCO

A year ago, I received a desperate call from the owner of an animal shelter near Munich. Local authorities conducting a raid had confiscated numerous fighting roosters as well as two fighting dogs. It was the conclusion of an investigation into illicit gambling, an underworld that may be utterly alien to you or me but that seems to lurk around every corner nonetheless. Animal fights leading to excruciating death among many of the animals involved is just one of its faces. Fighting roosters and fighting dogs present an enormous financial value to many a gambler. A lot of money is at stake for those involved, which is why they even sent people to break into the animal shelter in an attempt to recapture their living assets.

The owner and I then decided to transfer the animals to safe places and to hide them from their tormentors. One of the dogs, Michael, apparently had his ears cut off with a carpet knife. That's how these thugs try to make dogs more aggressive and to get them to fight longer. The dog with the mutilated ears is still under my personal protection today, at a location known only to me. The other dog was placed in a good home.

Gina, who is now a resident of Aiderbichl, also had her ears cut off by some thugs. That was in Bulgaria where she's from. Maybe it was done for kicks or for the same reason that the fighting dog lost his. Perhaps it was just some amateur attempting to trim her ears, nobody knows. But if there's anything these sadists could not take away from her, it's her charming, lovable nature that doesn't go unnoticed among the other dogs either.

Her best buddy is Arco, who's also from Bulgaria. Arco is scared to death of all people, except for us and his caretaker. We've made sure that he's going to remain with us, where he's safe from strangers.

Arthur und sein kurzes Glück

Es war ein ganz besonders armer Hund, den wir in einem Tierheim auf der griechischen Insel Thassos entdeckten. Er war vor einigen Wochen entkräftet am Strand gefunden worden. Die Tierschützer nahmen ihn trotz hoffnungsloser Überfüllung auf, versorgten ihn, ließen ihn ärztlich behandeln und mit entsprechenden Impfungen versehen. Doch Arthur ging es von Tag zu Tag schlechter. Die Tierklinik vor Ort hatte getan, was sie konnte, und so hofften die Helfer auf ein Wunder. Vielleicht nimmt ihn irgendjemand mit nach Deutschland in eine große Tierklinik – das schien der letzte Ausweg.

Das Wunder sollte geschehen. Vorsichtig legten wir Arthur in eine große Jet-Box, brachten ihn zum Flughafen und dann, sofort nach unserer Ankunft, in eine große Tierklinik nach München. Er vertraute uns und war dankbar, wenn wir ihn streichelten. An seinem Atem konnte man erkennen, dass er unsere Bemühungen wahrnahm und keine Angst hatte.

Nach einer langen Untersuchung kamen die Ärzte mit ernster Miene zu uns in den Warteraum. Dann wurde zur Darstellung einer Röntgenaufnahme eine Lichttafel eingeschaltet. Die Diagnose hätte nicht schrecklicher sein können: Auf Arthur hatten Menschen mit Schrotkugeln geschossen. Insgesamt steckten 50 Kugeln in seinem Körper. Aber nicht genug. Auch das Projektil eines Luftgewehrs war auf dem Röntgenbild zu sehen. Was muss dieser arme Hund mitgemacht haben? Wie gnadenlos sind die Menschen mit ihm umgegangen? Und was für ein Wunder, dass er überhaupt noch lebte. Die folgenden Tage waren entscheidend für ihn, aber Arthur sollte es nicht schaffen.

Arthurs Geschichte hat – im Gegensatz zu allen anderen in diesem Buch – kein Happy End. Doch irgendwie hat Arthur ganz zum Schluss doch noch etwas erfahren, das ihn vielleicht mit dieser Welt ein bisschen versöhnen konnte: Wir haben uns bei ihm entschuldigt und ihm gezeigt, dass für uns sein Leben von größter Bedeutung ist. Dass er es wert ist, umsorgt zu werden und dass uns für ihn keine Mühe zu groß ist. Denn Arthur war ein Lebewesen und ein Teil der Schöpfung, wie wir selber auch.

Arthur and his Brief Look at Happiness

There was a terribly poor dog that we came across at an animal shelter on the Greek Island of Thassos. He'd been discovered, at the end of his strength, on the beach some weeks ago. Even in the face of hopeless overcrowding, the staff of the animal shelter took him in, fed him and gave him medical treatment along with the proper vaccinations. Still, Arthur's condition declined from one day to the next. With the local animal clinic having done the best it could, all his rescuers could do was hope for a miracle. Maybe somebody would be willing to take him to a large animal clinic in Germany—it seemed like the only chance left.

Indeed, that miracle came to be. Carefully, we accommodated Arthur in a large jet box, took him to the airport and, immediately on our arrival, to a major animal clinic in Munich. He trusted us and was grateful when we stroked him. You could tell from his breathing that he appreciated our efforts and that he was unafraid.

Following a long examination, the doctors met us in the waiting room, their expressions looking grim. They proceeded to use an illuminating board to show us an x-ray. The diagnosis heightened our worst fears: Somebody had fired a buckshot at Arthur. He had no less than 50 pellets in his body. And that wasn't all, the x-ray also revealed the projectile of a BB gun. God only knew what must that poor dog have been through and how merciless some human had been on him! It was amazing that he was alive at all. The following days, his life was hanging by a thread. Sadly, Arthur didn't make it.

Unlike all the other stories in this book, Arthur's story does not have a happy end. Yet, Arthur somehow had the chance to experience something that might allow him to make at least some little peace with this world: We gave him an apology and let him feel that his life meant the world to us, that he was worth being cared for, and that we wouldn't spare any effort to do so. After all, Arthur was a living creature and as much a part of God's creation as we were.

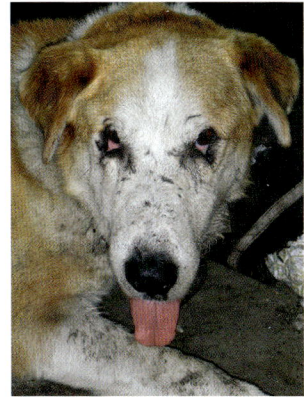

Was ich sonst noch über Hunde zu sagen hätte

What Else I'd Like to Add on the Subject of Dogs

WAS ICH SONST NOCH ÜBER HUNDE ZU SAGEN HÄTTE

Ich bin unseren Tierpflegern, besonders Hans Eder, zu unendlichem Dank verpflichtet. Ohne ihre Hilfe könnte ich überhaupt kein Haustier haben, denn Tiere brauchen Zeit. Aber dank der Unterstützung konnte ich besonders in den letzten 20 Jahren mein Wissen über die Besonderheiten der Tiere unglaublich vergrößern. Dankbar verneige ich mich vor all den lieben Menschen, die mir das ermöglicht haben.

So tun sich Mensch und Hund wechselseitig gut

Wir sind auf Aiderbichl in der glücklichen Lage, helfen zu können – auch dank vieler Mitstreiter, die uns mit Rat, Tat und auch finanziell unterstützen.

Aber nicht nur Menschen helfen Hunden, sondern auch Hunde helfen Menschen. Ich bin Vielen begegnet, die ihre Lebensqualität im Wesentlichen der Partnerschaft mit einem

WHAT ELSE I'D LIKE TO ADD ON THE SUBJECT OF DOGS

I owe my eternal gratitude to our animal caretakers and especially to Hans Eder. In fact, without their help, I wouldn't be able to keep any pets at all, because animals take time. Thanks to all the support, however, I've been able to gain tremendous insight into the personalities of these animals, especially within the last 20 years. My humble gratitude goes out to all these fantastic individuals, who made that possible for me.

How Man and Dog Can Benefit from Each Other

Here at Aiderbichl, we're in that lucky position enabling us to provide help, but the credit also goes to our many supporters, who assist us in word and deed, and financially as well.

However, it's not just people helping dogs, it's also dogs helping people. I've met many people, who owe the quality of their lives essentially to their partnership with a pet. Just think of all those countless senior citizens blessed with the fortune of keeping an animal. Or all those who lack social ties in their lives. Taking a dog out for a walk is a great way to meet other people. It's true what they say about the "flirt factor" of owning a dog and how it helps lonely souls find others.

Tier verdanken. Denken Sie doch nur an die zahllosen älteren Menschen, deren Glück es ist, sich um ein Tier sorgen zu können. Oder diejenigen, denen es an sozialer Bindung fehlt. Beim Hundespaziergang lernt man sich kennen und nicht umsonst spricht man von einem gewissen „Flirtfaktor" bei Hunden, der einsamen Menschen wieder zu Kontakten verhilft.

Leider gerät oft in Vergessenheit, welch wichtige Rolle Lawinenhunde spielen, nach einem Tornado, einem Erdbeben oder eben bei Lawinenabgängen. In solchen Fällen setzen wir dann unsere letzte Hoffnung auf die Hunde. Von Partnerhunden, Blindenhunden, Rettungshunden und Hunden im Dienst der Polizei und der Security ganz zu schweigen. Wenn man dann hört, dass Hunde Krebsgeschwüre besser als unsere technischen Geräte entdecken können, regt das zum Nachdenken an: Tiere verfügen über ausgeprägte Sinne. Sinne, die uns nützen und helfen. Sie gehören Lebewesen, die gleichermaßen Angst, Freude und Liebe fühlen.

Wie werde ich zum Hunderetter?

Da gibt es viel zu bedenken. Besonders das Selbstverständlichste, das wir ganz schnell übersehen: Wer A sagt, muss auch B sagen können. Das heißt: Die Rettung ist ja nur der Anfang.

Am besten, Sie testen sich gleich vor Ort im Tierheim. Wenn beispielsweise ein alter Schäferhund auf ein Zuhause wartet, fragen Sie sich, ob Sie ihn mitnehmen würden. Da muss nämlich auch mal das Ego ganz unzeitgemäß außer Acht gelassen werden.

Vor drei Jahren habe ich so eine Hündin aus dem Grazer Tierheim „Arche Noah" bei mir aufgenommen. Susi hatte nicht mehr viel Zeit, aber wir haben ihr letztes Jahr nach bestem Wissen und Gewissen jeden Tag mit vielen Momenten des Glücks noch lebenswert gemacht.

Trotz allem ist zu bedenken: Ein Tier mit ungewisser Vorgeschichte kann

Isn't it sad how we often tend to forget the vital role of search dogs in, say, the aftermath of a tornado, an earthquake or an avalanche? It is situations like those in which we place our last hope on these dogs. And let's not even get started on all those assistance dogs, guide dogs, rescue dogs, patrol dogs, and K9 Units. Then, when you factor in studies in which dogs are better at detecting cancer than our technical devices are, it really makes you think, doesn't it? Animals have very distinctive senses. Senses that we benefit from. And they belong to creatures that experience fear, joy and love just like we do.

How Can I Become a Dog Rescuer?

There's a lot to consider. In particular, the most obvious being what we tend to overlook all too easily. In for a penny, in for a pound. In other words, rescuing a dog is merely the beginning.

The best way would be to put yourself to the test by visiting your nearest animal shelter. For example, let's say you see an old shepherd dog waiting for a home. Ask yourself if you'd be willing to take him home. You see, if there ever was a time when you had to check your ego at the door, this is it.

That's how I adopted a female dog from the "Arche Noah" animal shelter in Graz, Austria, three years ago. Susi didn't have much time left, but we spent every day

problematisch sein. Es braucht Retter, die zu außergewöhnlichen Opfern bereit sind. Sonst verlieren die Hunde gleich wieder ihren Platz und noch ein Stück Vertrauen zum Menschen.

Wir leben in einer Zeit extremer Kontrolle. Das dürfen Sie nicht vergessen, wenn Sie einen Hund im Ausland, im Urlaub an sich gewöhnen, mit dem Ziel ihn mitzunehmen. Die Behörden haben mittlerweile Gesetze geschaffen, die es fast unmöglich machen, einen Hund im Ausland aufzunehmen und mit nach Hause zu bringen. Lesen Sie sich ganz genau durch, wie die Vorschriften lauten. Vertrauen Sie nicht darauf, dass Ihnen Verständnis und Toleranz bei der Einreise entgegengebracht werden. Nicht selten passiert es, dass Sie den Hund in sein Ursprungsland zurückschicken müssen oder unglaublich hohe Kosten auf Sie zukommen, wenn das Tier nach der Ankunft in Quarantäne kommt. Trotzdem bin ich der Meinung, dass sich alle Mühen lohnen. Setzen Sie sich im Urlaubsland mit Tierschützern in Verbindung, sie können Ihnen helfen.

doing everything in our power to make her last year worth living by giving her lots of moments full of happiness.

In spite of it all, let's not forget that a dog with an uncertain past can be a problem. As a rescuer, you have to be willing to make extraordinary sacrifices. Otherwise, that dog will lose yet another place and yet more confidence in mankind.

We live in a time of extreme control. Please remember that in case you're vacationing abroad and you decide to befriend some local dog with the intent of taking him home. These days, officials have introduced laws that make it almost impossible to pick up a dog in a foreign country and to take him home with you. Be sure to read all relevant regulations in detail. Don't ever expect compassion and tolerance upon your arrival. It's not rare for people having to send dogs back to their original countries or to face hideous costs when an animal is quarantined after its arrival. But I still hold the belief that every effort is worth it. In whatever country you choose, contact local animal protectionists, they can assist you.

Wenn Hunde gehen – Haben Tiere eine Seele?

Als ein befreundeter Journalist einen Bauern fragte, ob Tiere eine Seele haben, antwortete der Landwirt mit Metzgerei: „Wenn's richtig ist, muss es so sein. Das hier kann's doch nicht gewesen sein!" Angesichts der Erlebnisse, die ich in den letzten 20 Jahren mit Hundeschicksalen hatte, kann ich nur hoffen, dass er Recht behält.

Wenn der treue Freund, den man einschläfern musste, tot vor einem liegt, will man die Erinnerung an ihn ganz nahe bei sich behalten und würde ihn am liebsten in einem stillen Teil des Gartens begraben, was allerdings aus Gründen der Hygiene verboten ist. Es bleibt die Einäscherung, eine Urne, und für sie ein schöner Platz. Natürlich müssen wir zuerst an das Leben denken und dafür Sorge tragen. Aber es gibt keine Gegenwart ohne Vergangenheit. Erinnern und nicht vergessen, schon gar nicht einen Freund, das gehört zu unserer Kultur. Aber jenseits von Urnen können wir alle verstorbenen Tiere dadurch ehren, dass wir für ihre lebenden, leidenden und hilfesuchenden Artgenossen eintreten. Das nenne ich Respekt. Vor den Lebenden genauso wie vor den Toten.

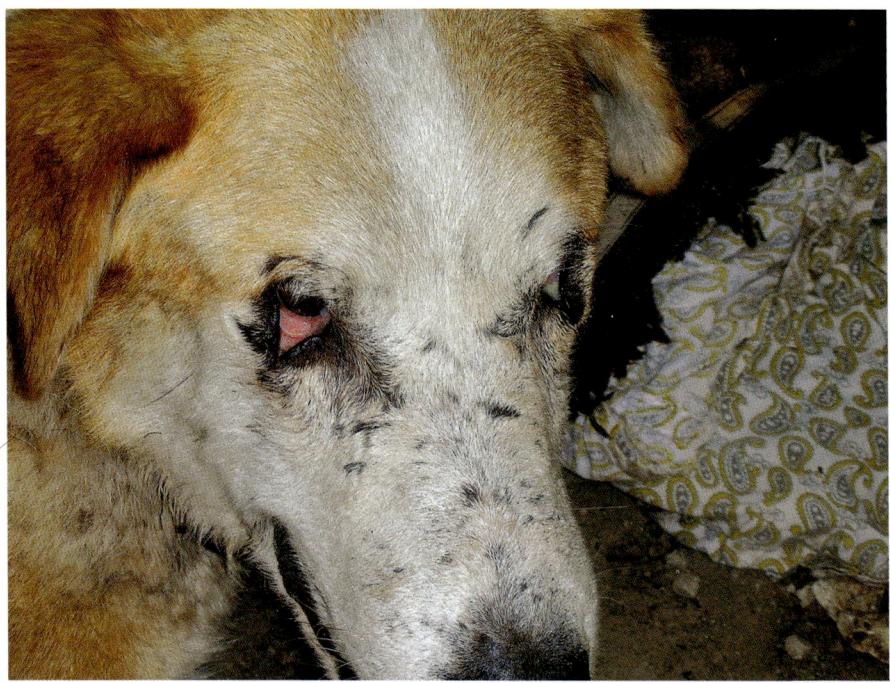

When Dogs Pass Away, Do Animals Have Souls?

When a friend of mine, a journalist, asked a farmer whether animals have souls, the farmer who also ran a butcher shop, said, "Well if they're meant to, then they've got to. Heck, it can't just end here!" Given all the things I've seen in canine fates within the last 20 years, I can only hope that he's right.

When man's best friend had to be put to sleep and you see his corpse, all you want to do is keep his memory as close to you as possible by burying him in a quiet place somewhere in your garden. Of course, you can't do that because of sanitary reasons. That leaves cremation, an urn and a beautiful place for it. Naturally, our first concern should be for life and we should act accordingly. Still, there's no present without a past. To remember and not forget is inherent to our culture, especially when it comes to a dear friend. Urns aside, however, the best way we can honor any deceased pet is by standing up for the other members of their species who are still living, suffering and in the need of help. That's what I call respect. Respect for all creatures, living and dead.

PATENSCHAFTEN

Zur Philosophie von Gut Aiderbichl: Es waren die Menschen, die sich für Gut Aiderbichl engagieren, die einen Begriff geprägt haben, der in drei Worten ihre Gesinnung beschreibt: „Ich bin Aiderbichler!" Mit Gut Aiderbichl haben sie einen Ort und Menschen gefunden, die ihre Art von Tierliebe, ihre Gedanken, Gefühle und Hoffnungen verstehen. Und deshalb haben sie sich entschlossen, „Aiderbichler" zu werden. Über 1000 Tiere stehen derzeit auf elf Höfen in Deutschland und Österreich unter unserem Schutz.

Wie kann ich „Aiderbichler" werden und die Anliegen von Gut Aiderbichl unterstützen?

Pate/Mitglied werden: Zum Beispiel mit einer Paten-/Mitgliedschaft, die schon ab €10,00 monatlich möglich ist und Ihnen und Ihren Begleitpersonen an 365 Tagen im Jahr nicht nur freien Eintritt auf unseren Gütern in Henndorf und Deggendorf bietet, sondern auch Zugriff auf unsere Live-Kameras im Internet und vieles mehr.

Hundepatenschaften: Die Haltung unserer Hunde ist auf einen lebenslangen Verbleib bei uns ausgerichtet. Unsere Hunde leben in unseren Privathaushalten, selbstverständlich ohne Zwingerhaltung. Schon lange planen wir ein Dorf für Hunde. Individuelle Häuser, in denen ein Pfleger mit einer Hundegruppe lebt. Bei der Haltung von Hunden legen wir größten Wert auf Erfüllung ihrer täglichen Bedürfnisse, eine zuverlässige Routine und die Möglichkeit, individuell wahrgenommen und geliebt zu werden. Beste medizinische Versorgung und ein abwechslungsreiches tägliches Bewegungsprogramm wird allen unseren Hunden garantiert. Hundepatenschaften können auf Wunsch mit der Benennung eines einzelnen Tieres, das bei uns lebt, abgeschlossen werden. Die Einnahmen kommen allen geretteten Hunden zugute.

Förderer der Gut Aiderbichl Stiftung werden: Außer den beiden großen bekannten Gütern gibt es noch neun weitere Höfe, auf denen viele von uns gerettete Tiere leben. Sie werden ausschließlich von den gemeinnützigen Gut Aiderbichl Stiftungen in Deutschland und Österreich unterstützt. Bitte helfen Sie uns. Spenden Sie und werden Sie schon mit einer einmaligen Spende Förderer von Gut Aiderbichl.

Um „Aiderbichler" zu werden, wenden Sie sich bitte an: Gut Aiderbichl Verwaltung, Johannes-Filzer-Straße 5, 5020 Salzburg oder telefonisch an: +43 (662) 62 53 95 oder per E-Mail an: info@gut-aiderbichl.com.

Sponsorships

About the Philosophy of Gut Aiderbichl: It was the people dedicating themselves to Gut Aiderbichl, who coined a phrase that puts their philosophy into three words, "I am an Aiderbichler!" In Gut Aiderbichl, they have found a place and people that identify with their level of love for animals, their thoughts, feelings, and hopes. That's what made them decide to become an "Aiderbichler." Currently, we have more than 1,000 animals under our care and protection in eleven sanctuaries in Germany and Austria.

How can I become an "Aiderbichler" and support the cause of Gut Aiderbichl?

Becoming a Sponsor/Member: A sponsorship or membership, for example, is available starting at just €10.00 ($15) a month. Not only does it provide you and traveling companions free admission to our sanctuaries in Henndorf and Deggendorf, but also with access to our live cameras on the Internet and a whole lot more.

Sponsoring Dogs: Our dog care is geared towards the idea that they can spend the rest of their lives with us. We keep our dogs in our private households. Kennels are something we can do without. For a long time, we've been planning a village for dogs. It involves individual houses each accommodating one caretaker with a group of dogs. Our dog care centers on meeting their daily needs, establishing a routine life for them and to accept and love them for the individuals they are. We also guarantee top-quality medical supplies and a daily flexible exercise program for all our dogs. On request, sponsorships of dogs can be arranged just by naming an individual animal that lives with us. Proceeds are used for the benefit of all rescued dogs.

Becoming a Promoter of the Gut Aiderbichl Foundation: In addition to our two major and well-known sanctuaries, we have nine more sanctuaries sheltering many animals we've rescued. They are exclusively supported by our non-profit Gut Aiderbichl Foundations in Germany and Austria. Please help us. By making just one donation, you can become a promoter of Gut Aiderbichl.

To become an "Aiderbichler," please contact: Gut Aiderbichl Verwaltung, Johannes-Filzer-Straße 5, 5020 Salzburg, Austria, phone +43 (662) 62 53 95, or send your e-mail to: info@gut-aiderbichl.com.

IMPRINT

© 2008 teNeues Verlag GmbH + Co. KG, Kempen
Text and photographs © Gut Aiderbichl GmbH
All rights reserved.

Editor: Michael Aufhauser
Gut Aiderbichl Stiftung, Gut Aiderbichl GmbH, Johannes-Filzer-Str. 5, A-5020 Salzburg
Responsible for content: Michael Aufhauser, Dieter Ehrengruber, Friederike Grünthal
Associates: Christian Dutz, Michaela Kalss, Sabine Schlömer, Helmut Schödel, Holde Sudenn
Photographs by Dieter Ehrengruber, Andreas Kolarik, Franz-Josef Lang, Franz Neumayr, Agnes
Schindler, Alexandra Schlump, Markus Tschepp, Jürgen Weyrich, Zeppelzauer
Translation by Artes Translations: Conan Kirkpatrick
Design by Robert Kuhlendahl, Iris Durie
Production by Sandra Jansen
Editorial coordination by Pit Pauen
Color separation by MT-Vreden, Vreden

Published by teNeues Publishing Group

teNeues Verlag GmbH + Co. KG
Am Selder 37
47906 Kempen, Germany
Tel.: 0049-(0)2152-916-0
Fax: 0049-(0)2152-916-111
e-mail: books@teneues.de

Press department: Andrea Rehn
Tel.: 0049-(0)2152-916-202
e-mail: arehn@teneues.de

www.teneues.com

teNeues Publishing Company
16 West 22nd Street
New York, NY 10010, USA
Tel.: 001-212-627-9090
Fax: 001-212-627-9511

teNeues Publishing UK Ltd.
P.O. Box 402
West Byfleet
KT14 7ZF, Great Britain
Tel.: 0044-1932-4035-09
Fax: 0044-1932-4035-14

teNeues France S.A.R.L.
93, rue Bannier
45000 Orléans, France
Tel.: 0033-2-3854-1071
Fax: 0033-2-3862-5340

ISBN: 978-3-8327-9278-7

Printed in Italy

teNeues Publishing Group
Kempen
Düsseldorf
Hamburg
London
Munich
New York
Paris

teNeues